Principles of Molecular Virology

science &
technology books

ELSEVIER

Companion Web Site:

www.store.elsevier.com/9780128019467

Principles of Molecular Virology
Alan J. Cann

Resources available:

- All figures from the book available as PPT slides and .jpeg files
- Glossary from the book
- Self-assessment questions

ELSEVIER

TOOLS FOR ALL YOUR TEACHING NEEDS
textbooks.elsevier.com

ACADEMIC
PRESS

Principles of Molecular Virology

Sixth Edition

Alan J. Cann

Department of Biology,
University of Leicester,
Leicester, UK

AMSTERDAM • BOSTON • HEIDELBERG • LONDON
NEW YORK • OXFORD • PARIS • SAN DIEGO
SAN FRANCISCO • SINGAPORE • SYDNEY • TOKYO

Academic Press is an imprint of Elsevier

Academic Press is an imprint of Elsevier
125 London Wall, London EC2Y 5AS
525 B Street, Suite 1800, San Diego, CA 92101-4495, USA
225 Wyman Street, Waltham, MA 02451, USA
The Boulevard, Langford Lane, Kidlington, Oxford OX5 1GB, UK

Sixth Edition

Notices
Knowledge and best practice in this field are constantly changing. As new research and experience broaden our understanding, changes in research methods, professional practices, or medical treatment may become necessary.

Practitioners and researchers must always rely on their own experience and knowledge in evaluating and using any information, methods, compounds, or experiments described herein. In using such information or methods they should be mindful of their own safety and the safety of others, including parties for whom they have a professional responsibility.

To the fullest extent of the law, neither the Publisher nor the authors, contributors, or editors, assume any liability for any injury and/or damage to persons or property as a matter of products liability, negligence or otherwise, or from any use or operation of any methods, products, instructions, or ideas contained in the material herein.

Library of Congress Cataloging-in-Publication Data
A catalog record for this book is available from the Library of Congress

British Library Cataloguing-in-Publication Data
A catalogue record for this book is available from the British Library

ISBN: 978-0-12-801946-7

For information on all Academic Press publications
visit our website at http://store.elsevier.com/

Typeset by MPS Limited, Chennai, India
www.adi-mps.com

Publisher: Janice Audet
Acquisition Editor: Jill Leonard
Editorial Project Manager: Halima Williams
Production Project Manager: Julia Haynes
Designer: Ines Cruz

Contents

CHAPTER 6

Preface to the Sixth Edition

In the age of the Internet, why would anyone write a textbook about virology? Indeed, why would anyone write anything about virology? Virology isn't dead yet (DiMaio, 2014), and neither are books. I encourage everyone to use the wonderful resource of the Internet to improve their knowledge of virology. I encourage my students to use Wikipedia and Google to learn the facts. But as Jimmy Wales said, *Wikipedia is often the best place to start, but the worst place to stop.* The role of this book is not primarily about knowledge but about sense-making—what you can't get from Wikipedia. Virology *explained* by setting facts in a larger context.

Along with updating the facts and smoothing some of the rough edges, I have noticed a big scientific change in writing this edition. Open Access scientific publishing has finally made its impact felt. In this updated edition the reading recommendations at the end of each chapter I have been able, in *almost* all cases, to recommend freely available peer-reviewed content for readers. You may have to hunt around to find it—a good working knowledge of PubMed and Google Scholar is at least as useful as Google and Wikipedia—but it is now possible to access much of the scientific literature the public has paid for. But there is still the question of interpretation. In writing this book I have tried to do my part. The rest is up to the reader.

As with previous editions, I am grateful to the staff of Elsevier, in particular Halima Williams and Jill Leonard, for their patience with me.

Alan J. Cann
University of Leicester, UK
alan.cann@leicester.ac.uk
December 2014

Reference

DiMaio, D., 2014. Is virology dead? mBio 5 (2), e01003–e01014.

CHAPTER 1

Introduction

Intended Learning Outcomes

On completing this chapter you should be able to:

- Define how viruses are different from other biological organisms.
- Explain how the development of virology led us to our present understanding of viruses.
- Be able to discuss how technology has influenced the study of viruses in recent years.

CONTENTS

This book is about "molecular virology," that is, the molecular basis of how viruses work. It looks at the protein–protein, protein–nucleic acid, and protein–lipid interactions which control the structure of virus particles, the ways viruses infect cells, and how viruses replicate themselves. Later we will also examine the consequences of virus infection for host organisms, but it is important to consider the basic nature of viruses first. To understand how our present knowledge of viruses was achieved, it will be useful to know a little about the history of virology. This helps to explain how we think about viruses and what the current and future concerns of virologists are.

There is more biological diversity between different viruses than in all the rest of the bacterial, plant, and animal kingdoms put together. This is the result of the success of viruses in parasitizing all known groups of living organisms, and understanding this diversity is the key to comprehending the interactions of viruses with their hosts. The principles behind some of the experimental techniques mentioned in this chapter may not be well known to all readers. That is why it may be helpful to you to use the further reading at the end of this chapter to become more familiar with these methods or you will not be able to understand the current research literature you read. In this and the subsequent chapters, terms in the text in **bold red print** are defined in the glossary at the end of the book (Appendix 1).

Principles of Molecular Virology. DOI: http://dx.doi.org/10.1016/B978-0-12-801946-7.00001-8

WHAT ARE VIRUSES?

Viruses are submicroscopic, obligate intracellular parasites. Most are too small to be seen by optical microscopes, and they have no choice but to replicate inside host cells. This simple but useful definition goes a long way toward describing viruses and differentiating them from all other types of organism. However, this short definition is not completely adequate. It is not a problem to differentiate viruses from multicellular organisms such as plants and animals. Even within the broad scope of microbiology, covering prokaryotic organisms as well as microscopic eukaryotes such as algae, protozoa, and fungi, in most cases this simple definition is enough. A few groups of prokaryotic organisms also have specialized intracellular parasitic life cycles and overlap with this description. These are the *Rickettsiae* and *Chlamydiae*-obligate intracellular parasitic bacteria which have evolved to be so cell-associated that they can exist outside the cells of their hosts for only a short period of time before losing viability.

A common mistake is to say that viruses are smaller than bacteria. While this is true in most cases, size alone does not distinguish them. The largest virus known (currently *Pithovirus sibericum*) is 1,200 nm long, while the smallest bacteria (e.g., *Mycoplasma*) are only 200−300 nm long. Nor does genetic complexity separate viruses from other organisms. The largest virus genome (Pandoravirus, 2.8 Mbp—million base pairs—approximately 2,500 genes) is twenty times as big as smallest bacterial genome (*Tremblaya princeps*, at 139 kbp—thousand base pairs—and with only 120 protein coding genes), although it is still shorter than the smallest eukaryotic genome (the parasitic protozoan *Encephalitozoon*, 2.3 Mbp). For these reasons, it is necessary to go further to produce a definition of how viruses are unique:

- Virus particles are produced from the assembly of preformed components, while other biological agents grow from an increase in the integrated sum of their components and reproduce by division.
- Virus particles (virions) do not grow or undergo division.
- Viruses lack the genetic information that encodes the tools necessary for the generation of metabolic energy or for protein synthesis (ribosomes).

No known virus has the biochemical or genetic means to generate the energy necessary to drive all biological processes. They are absolutely dependent on their host cells for this function. Lacking the ability to make ribosomes is one factor which clearly distinguishes viruses from all other organisms. Although there will always be some exceptions and uncertainties in the case of organisms that are too small to see easily and in many cases difficult to study, the above guidelines are sufficient to define what a virus is.

A number of virus-like agents possess properties that confuse the above definition yet are clearly more similar to viruses than other organisms. These are the subviral elements known as viroids, virusoids, and prions. Viroids are small (200–400 nucleotide), circular RNA molecules with a rod-like secondary structure. They have no capsid or envelope and are associated with certain plant diseases. Their replication strategy is like that of viruses—they are obligate intracellular parasites. Virusoids are satellite, viroid-like molecules, a bit larger than viroids (approximately 1,000 nucleotides), which are dependent on the presence of virus replication for their multiplication (the reason they are called "satellites"). They are packaged into virus capsids as passengers. Prions are infectious protein molecules with no nucleic acid component. Confusion arises from the fact that the prion protein and the gene that encodes it are also found in normal "uninfected" cells. These agents are associated with diseases such as Creutzfeldt–Jakob disease in humans, scrapie in sheep, and bovine spongiform encephalopathy (BSE) in cattle. Chapter 8 deals with these subviral infectious agents in more detail.

Genome analysis has shown that more than 10% of the eukaryotic cell genome is composed of mobile retrovirus-like elements (retrotransposons), which may have had a considerable role in shaping these complex genomes (Chapter 3). Furthermore, certain bacteriophage genomes closely resemble bacterial plasmids in their structure and in the way they are replicated. Research has revealed that the evolutionary relationship between viruses and other living organisms is perhaps more complex than was previously thought.

ARE VIRUSES ALIVE?

As discussed earlier, viruses do not reproduce by division but are assembled from preformed components, and they cannot make their own energy or proteins. A virus-infected cell is more like a factory than a womb. One view is that inside their host cell viruses are alive, whereas outside it they are only complex arrangements of metabolically inert chemicals. Chemical changes do occur in extracellular virus particles, as explained in Chapter 4, but these are not in the "growth" of a living organism. This is a bit problematic—alive at sometimes but not at others. Viruses do not fit into most of the common definitions of "life"—growth, respiration, etc. Ultimately, whether viruses are alive or not is a matter of personal opinion, but it is useful to make your decision after considering the facts. Some of the reading at the end of this chapter will help you consider the evidence.

BOX 1.1 ARE VIRUSES ALIVE? WHO CARES?

Viruses don't care (can't care) if we think they are living or not. And I don't care much either, because as far as I'm concerned it is much more important to understand how viruses replicate themselves and interact with their hosts. But you might care, either because you are a philosophical person who likes thinking about these things, or because you have to write an essay or answer an exam question on the subject. In that case, it is important to consider how you define what a living organism is and how viruses are similar or different to microorganisms we consider to be alive (you're going to make life hard for yourself if you start comparing them to humans). This is not a simple question, and any simple answer is, quite simply, wrong.

THE HISTORY OF VIROLOGY

Human knowledge of virus diseases goes back a long way, although it is only much more recently that we have become aware of viruses as distinct from other causes of disease. The first written record of a virus infection is a hieroglyph from Memphis, the capital of ancient Egypt, drawn in approximately 3700 BC, which depicts a temple priest showing typical clinical signs of paralytic poliomyelitis. Pharaoh Ramses V, who died in 1196 BC and whose well-preserved mummified body is now in a Cairo museum, is believed to have died from smallpox—the comparison between the pustules on the face of this mummy and those of more recent patients is startling.

Smallpox was **endemic** in China by 1000 BC. In response, the practice of **variolation** was developed. Recognizing that survivors of smallpox outbreaks were protected from subsequent infection, people inhaled the dried crusts from smallpox lesions like snuff or, in later modifications, inoculated the pus from a lesion into a scratch on the forearm. Variolation was practiced for centuries and was shown to be an effective method of disease prevention, although risky because the outcome of the inoculation was never certain. Edward Jenner was nearly killed by variolation at the age of seven. Not surprisingly, this experience spurred him on to find a safer alternative treatment. On May 14, 1796, he used cowpox-infected material obtained from the hand of Sarah Nemes, a milkmaid from his home village of Berkeley in Gloucestershire, England, to successfully vaccinate 8-year-old James Phipps. Although initially controversial, **vaccination** against smallpox was almost universally adopted worldwide during the nineteenth century.

This early success, although a triumph of scientific observation and reasoning, was not based on any real understanding of the nature of infectious agents. This arose separately from another line of reasoning. Antony van Leeuwenhoek (1632−1723), a Dutch merchant, constructed the first simple microscopes and with these identified bacteria as the "animalcules" he saw in his specimens. However, it was not until Robert Koch and Louis Pasteur in the 1880s jointly proposed the "germ theory" of disease that the significance of these organisms became apparent. Koch defined four famous criteria which are now known as Koch's postulates and still generally regarded as the proof that an infectious agent is responsible for a specific disease:

1. The agent must be present in every case of the disease.
2. The agent must be isolated from the host and grown *in vitro*.
3. The disease must be reproduced when a pure culture of the agent is inoculated into a healthy susceptible host.
4. The same agent must be recovered once again from the experimentally infected host.

Subsequently, Pasteur worked extensively on rabies, which he identified as being caused by a "virus" (from the Latin for "poison"), but despite this he did not discriminate between bacteria and other agents of disease. In 1892, Dimitri Iwanowski, a Russian botanist, showed that extracts from diseased tobacco plants could transmit disease to other plants after being passed through ceramic filters fine enough to retain the smallest known bacteria. Unfortunately, he did not realize the full significance of these results. A few years later (1898), Martinus Beijerinick confirmed and extended Iwanowski's results on tobacco mosaic virus (TMV) and was the first to develop the modern idea of the virus, which he referred to as *contagium vivum fluidum* ("soluble living germ"). Freidrich Loeffler and Paul Frosch (1898) showed that a similar agent was responsible for foot-and-mouth disease in cattle, but, despite the realization that these new-found agents caused disease in animals as well as plants, people would not accept the idea that they might have anything to do with human diseases. This resistance was finally dispelled in 1909 by Karl Landsteiner and Erwin Popper, who showed that poliomyelitis was caused by a "filterable agent"—the first human disease to be recognized as being caused by a virus.

Frederick Twort (1915) and Felix d'Herelle (1917) were the first to recognize viruses that infect bacteria, which d'Herelle called bacteriophages ("eaters of bacteria"). In the 1930s and subsequent decades, pioneering virologists such as Salvador Luria, Max Delbruck, and others used these viruses as model systems to investigate many aspects of virology, including virus structure (Chapter 2), genetics (Chapter 3), and replication (Chapter 4). These relatively simple agents have since proved to be very important to our understanding of all types of viruses, including those of humans which can be much more difficult to propagate and study. The further history of virology is the story of the development of experimental tools and systems with which viruses could be examined and which opened up whole new areas of biology, including not only the biology of the viruses themselves but inevitably also the biology of the host cells on which they are dependent.

LIVING HOST SYSTEMS

In 1881, Louis Pasteur began to study rabies in animals. Over several years, he developed methods of producing **attenuated** virus preparations by progressively drying the spinal cords of rabbits experimentally infected with rabies which, when inoculated into other animals, would protect from disease caused by virulent rabies virus. In 1885, he inoculated a child, Joseph Meister, with this, the first artificially produced virus **vaccine** (since the ancient practice of **variolation** and Jenner's use of cowpox virus for **vaccination** had relied on naturally occurring viruses). Whole plants have been used to study the effects of plant viruses after infection ever since TMV was first discovered by Iwanowski in 1892. Usually such studies involve rubbing preparations containing virus particles into the leaves or stem of the plant to cause infection.

During the Spanish–American War of the late nineteenth century and the subsequent building of the Panama Canal, the number of American deaths due to yellow fever was colossal. The disease also appeared to be spreading slowly northward into the continental United States. In 1900, through experimental transmission of the disease to mice, Walter Reed demonstrated that yellow fever was caused by a virus spread by mosquitoes. This discovery eventually enabled Max Theiler in 1937 to propagate the virus in chick embryos and to produce an attenuated vaccine—the 17D strain—which is still in use today. The success of this approach led many other investigators from the 1930s to the 1950s to develop animal systems to identify and propagate pathogenic viruses.

Cultures of **eukaryotic** cells can be grown in the laboratory and viruses can be propagated in these cultures, but these techniques are expensive and technically demanding. Some viruses such as influenza virus will replicate in the living tissues of developing embryonated hens' eggs. Egg-adapted strains of influenza virus replicate well in eggs and very high virus **titers** can be obtained. Embryonated hens' eggs were first used to propagate viruses in the early decades of the twentieth century. This method proved to be highly effective for the isolation and culture of many viruses, particularly strains of influenza virus and various poxviruses (e.g., vaccinia virus). Counting the "pocks" on the chorioallantoic membrane of eggs produced by the replication of vaccinia virus was the first quantitative assay for any virus. Animal host systems still have their uses in virology:

- To produce viruses that cannot be effectively studied *in vitro* (e.g., hepatitis B virus).
- To study the pathogenesis of virus infections (e.g. human immunodeficiency virus, HIV, and its near relative, simian immunodeficiency virus, SIV).
- To test vaccine safety (e.g., oral poliovirus vaccine).

Nevertheless, they are increasingly being discarded for the following reasons:

- Breeding and maintenance of animals infected with pathogenic viruses is expensive.
- Animals are complex systems in which it is sometimes difficult to isolate the effects of virus infection.
- Results obtained are not always reproducible due to host variation.
- Unnecessary or wasteful use of experimental animals is morally unacceptable.

With the exception of studying pathogenesis, the use of animals is generally being overtaken by molecular biology methods which are faster and cheaper. In the 1980s the first transgenic animals were produced which carried the genes of other organisms. Inserting all or part of a virus genome into the DNA of an embryo (typically of a mouse) results in expression of virus mRNA and proteins in the animal. This allows the pathogenic effects of virus proteins, individually and in various combinations, to be studied in living hosts. "Humanized" mice have been constructed from immunodeficient animals transplanted with human tissue. These mice form an intriguing model to study the pathogenesis of HIV as there is no real alternative to study the properties of HIV *in vivo*. Similarly, transgenic mice have proved to be vitally important in understanding the biology of prion genes. While these techniques raise the same moral objections as "old-fashioned" experimental infection of animals by viruses, they are immensely powerful new tools for the study of virus pathogenicity. A growing number of plant and animal viruses genes have been analyzed in this way, but the results have not always been as expected, and in some cases it has proved difficult to equate the observations obtained with those gathered from experimental infections. Nevertheless, this method has become quite widely used in the study of important diseases where few alternative models exist.

BOX 1.2 WHAT'S THE PROBLEM WITH TRANSGENICS?

For thousands of years farmers have transferred genes from one species of plant into another by crossing two or more species. This is the way that wheat was created over 10,000 years ago. There was no control, other than trial and error, over which genes were transferred or the properties the resulting offspring possessed. In the 1980s it became possible to genetically modify plants and animals by transferring specific genes or groups of genes from another species. And so the controversy over GM crops arose—were they the saviors of humanity, feeding the starving and reducing pollution, or heralds of environmental doom? At about the same time, the first transgenic mice were made. Although there was an outcry at the time, this was dwarfed by the controversy over the first transgenic monkey in 2001. Genetically modified versions of our human relatives seemed too close to home for some people, reminding them of eugenics, the selective breeding of humans with its negative political and moral associations. In truth, science and technology are neutral, and it is societies who ultimately decide how they are used. Should we use these new technologies to feed the world and cure disease, or abandon them for fear of misuse? It's not the technology, it's what we do with it that matters.

CELL CULTURE METHODS

Cell culture began early in the twentieth century with whole-organ cultures, then progressed to methods involving individual cells, either **primary cell** cultures (cells from an experimental animal or taken from a human patient which can be maintained for a short period in culture) or **immortalized** cell lines, which, given appropriate conditions, continue to grow in culture indefinitely. In 1949, John Enders and his colleagues were able to propagate poliovirus in primary human cell cultures. In the 1950s and 1960s, this achievement led to the identification and isolation of many viruses and their association with human diseases—for example, many common viruses such as enteroviruses and adenoviruses. Widespread virus isolation led to the realization that subclinical virus infections with no obvious symptoms were very common; for example, even in **epidemics** of the most virulent strains of poliovirus there are approximately 100 subclinical infections for each paralytic case of poliomyelitis.

Renato Dulbecco in 1952 was the first to quantify accurately animal viruses using a **plaque** assay. In this technique, dilutions of the virus are used to infect a cultured cell **monolayer**, which is then covered with soft agar to restrict diffusion of the virus. This results in localized cell killing where a cell has been infected with the virus, and the appearance of plaques after the monolayer is stained (Figure 1.1). Counting the number of plaques measures the number of infectious virus particles applied to the plate. The same technique can also be used biologically to clone a virus (i.e., isolate a pure form from a mixture of types). This technique had been in use for some time to quantify the number of infectious virus particles in **bacteriophage** suspensions applied to confluent "lawns" of bacterial cells on agar plates, but its application to viruses of **eukaryotes** enabled rapid advances in the study of virus replication to be made. Plaque assays largely replaced earlier endpoint dilution techniques, such as the tissue culture infectious dose ($TCID_{50}$) assay, which are statistical means of measuring virus populations in culture. Endpoint techniques may still be used in certain circumstances—for example, for viruses that do not replicate in culture or are not cytopathic and do not produce plaques (e.g., HIV).

Virus infection has often been used to probe the working of "normal" (i.e., uninfected) cells—for example, to look at macromolecular synthesis. This is true of the applications of **bacteriophages** in bacterial genetics, and in many instances where the study of eukaryotic viruses has revealed fundamental information about the cell biology and genomic organization of higher organisms. Polyadenylation of messenger RNAs (mRNAs) (1970), chromatin structure (1973), and mRNA splicing (1977) were all discovered in viruses before it was realized that they could also be found in uninfected cells.

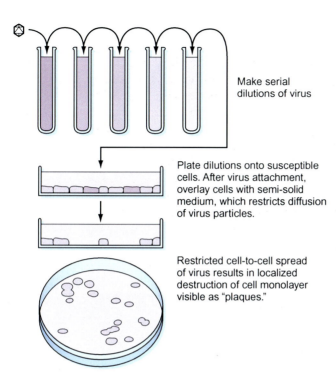

Make serial
dilutions of virus

Plate dilutions onto susceptible
cells. After virus attachment,
overlay cells with semi-solid
medium, which restricts diffusion
of virus particles.

Restricted cell-to-cell spread
of virus results in localized
destruction of cell monolayer
visible as "plaques."

FIGURE 1.1 Plaque assays.
Plaque assays are performed by adding a suitable dilution of a virus preparation to an adherent
monolayer of susceptible cells. After allowing time for virus attachment to the cells, a semi-solid culture
medium containing a polymer such as agarose or carboxymethyl cellulose, which restricts diffusion of
virus particles from infected cells, is added. Only direct cell-to-cell virus spread occurs, resulting in
localized destruction of cells. After incubation, the medium is removed and the cells stained to make the
holes in the monolayer (plaques) more visible. Each plaque results from infection by a single
plaque-forming unit (p.f.u.) allowing the original number of virus particles to be estimated
(but read the glossary definition of p.f.u. in Appendix 1).

SEROLOGICAL/IMMUNOLOGICAL METHODS

As the discipline of virology was emerging, the techniques of immunology
were also developing, and the two fields have always been very closely
linked. Understanding mechanisms of immunity to virus infections has, of
course, been very important. More recently, the role that the immune system
itself plays in pathogenesis has become known (see Chapter 7). Immunology
as a specialty has contributed many important techniques to virology
(Figure 1.2).

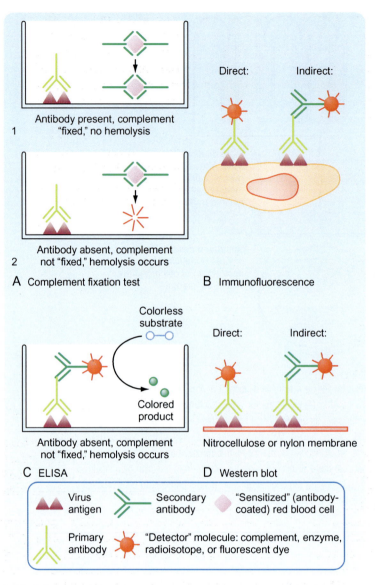

FIGURE 1.2 Serological techniques in virology.

The four assays illustrated in this figure have been used for many years and are of widespread value in many circumstances. They are used to test both viruses and for immune responses against virus infection.

(A) The *complement fixation test* works because complement is bound by antigen—antibody complexes. "Sensitized" (antibody-coated) red blood cells, known amounts of complement, a virus antigen, and the serum to be tested are all added to the wells of a multiwell plate. In the absence of antibodies to the virus antigen, free complement is present which causes lysis of the sensitized red blood cells (hemolysis). If the test serum contains a sufficiently high titer of antivirus antibodies, then no free complement remains and hemolysis

In 1941 George Hirst observed hemagglutination of red blood cells by influenza virus (see Chapter 4). This proved to be an important tool in the study of not only influenza but also several other groups of viruses—for example, rubella virus. In addition to measuring the titer (i.e., relative amount) of virus present in any preparation, this technique can also be used to determine the antigenic type of the virus. Hemagglutination will not occur in the presence of antibodies that bind to and block the virus hemagglutinin. If an antiserum is titrated against a given number of hemagglutinating units, the hemagglutination inhibition titer and specificity of the antiserum can be determined. Also, if antisera of known specificity are used to inhibit hemagglutination, the antigenic type of an unknown virus can be determined. In the 1960s and subsequent years, many improved detection methods for viruses were developed, such as:

- Complement fixation tests
- Radioimmunoassays
- Immunofluorescence (direct detection of virus antigens in infected cells or tissue)
- Enzyme-linked immunosorbent assays (ELISAs)
- Radioimmune precipitation
- Western blot assays

These techniques are sensitive, quick, and quantitative.

does not occur. Titrating the test serum through serial dilutions allows a quantitative measurement of the amount of antivirus antibody present to be made.

(B) *Immunofluorescence* is performed using modified antibodies linked to a fluorescent molecule that emits colored light when illuminated by light of a different wavelength. In direct immunofluorescence, the antivirus antibody is conjugated to the fluorescent marker, whereas in indirect immunofluorescence a second antibody reactive to the antivirus antibody carries the fluorescent marker. Immunofluorescence can be used not only to identify virus-infected cells in populations of cells or in tissue sections but also to determine the subcellular localization of particular virus proteins (e.g., in the nucleus or in the cytoplasm).

(C) *Enzyme-linked immunosorbent assays* (ELISAs) are a rapid and sensitive means of identifying or quantifying small amounts of virus antigens or antivirus antibodies. Either an antigen (in the case of an ELISA to detect antibodies) or antibody (in the case of an antigen ELISA) is bound to the surface of a multiwell plate. An antibody specific for the test antigen, which has been conjugated with an enzyme molecule (such as alkaline phosphatase or horseradish peroxidase), is then added. As with immunofluorescence, ELISA assays may rely on direct or indirect detection of the test antigen. During a short incubation, a colorless substrate for the enzyme is converted to a colored product, amplifying the signal produced by a very small amount of antigen. The intensity of the colored product can be measured in a specialized spectrophotometer ("plate reader"). ELISA assays can be mechanized and are suitable for routine tests on large numbers of clinical samples.

(D) *Western blotting* is used to analyze a specific virus protein from a complex mixture of antigens. Virus antigen-containing preparations (particles, infected cells, or clinical materials) are subjected to electrophoresis on a polyacrylamide gel. Proteins from the gel are then transferred to a nitrocellulose or nylon membrane and immobilized in their relative positions from the gel. Specific antigens are detected by allowing the membrane to react with antibodies directed against the antigen of interest. By using samples containing proteins of known sizes in known amounts, the apparent molecular weight and relative amounts of antigen in the test samples can be determined.

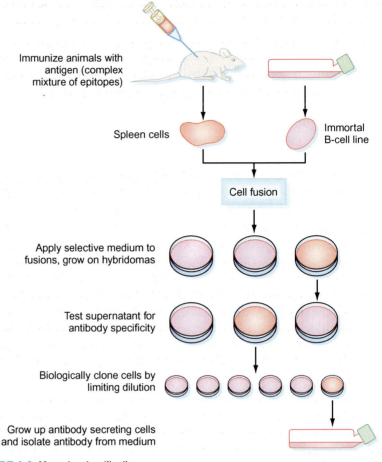

Immunize animals with
antigen (complex
mixture of epitopes)

Spleen cells

Immortal
B-cell line

Cell fusion

Apply selective medium to
fusions, grow on hybridomas

Test supernatant for
antibody specificity

Biologically clone cells by
limiting dilution

Grow up antibody secreting cells
and isolate antibody from medium

FIGURE 1.3 Monoclonal antibodies.
Monoclonal antibodies are produced by immunization of an animal with an antigen that usually contains
a complex mixture of epitopes. Immature B-cells are prepared from the spleen of the animal, and these
are fused with a myeloma cell line, resulting in the formation of transformed cells continuously secreting
antibodies. A small proportion of these will make a single type of antibody (a monoclonal antibody)
against the desired epitope. Recently, *in vitro* molecular techniques have been developed to speed up
the selection of monoclonal antibodies.

In 1975, George Kohler and Cesar Milstein isolated the first monoclonal antibodies from clones of cells selected *in vitro* to produce an antibody of a single
specificity directed against a particular antigen. This enabled virologists to
look not only at the whole virus, but at specific regions—epitopes—of individual virus antigens (Figure 1.3). This ability has greatly increased our understanding of the function of individual virus proteins. Monoclonal antibodies

found increasingly widespread application in other types of serological assays (e.g., ELISAs) to increase their reproducibility, sensitivity, and specificity.

It would be inappropriate here to devote too much discussion to the technical details of immunology, a very rapidly expanding field of knowledge. If you are not familiar with the techniques mentioned above, you should familiarize yourself with them by reading one of the many textbooks available.

ULTRASTRUCTURAL STUDIES

Ultrastructural studies can be considered under three areas: physical methods, chemical methods, and electron microscopy. Physical measurements of virus particles began in the 1930s with the earliest determinations of their proportions by filtration through membranes with various pore sizes. Experiments of this sort led to the first (rather inaccurate) estimates of the size of virus particles. The accuracy of these estimates was improved greatly by studies of the sedimentation properties of viruses in ultracentrifuges in the 1960s (Figure 1.4). Differential centrifugation was of great value in obtaining purified and highly concentrated preparations of many different viruses, free of contamination from host cell components, which could be subjected to chemical analysis. The relative density of virus particles, measured in sucrose or CsCl solutions, is also a characteristic feature, revealing information about the proportions of nucleic acid and protein in the particles.

Chemical investigation can be used to determine not only the overall composition of viruses and the nature of the nucleic acid that comprises the virus genome but also the construction of the particle and the way in which individual components relate to each other in the capsid. Many classic studies of virus structure have been based on the gradual, stepwise disruption of particles by slow alteration of pH or the gradual addition of protein-denaturing agents such as urea, phenol, or detergents. Under these conditions, valuable information can sometimes be obtained from relatively simple experiments. The reagents used to denature virus capsids can indicate the basis of the stable interactions between its components. Proteins bound together by electrostatic interactions can be eluted by addition of ionic salts or alteration of pH; those bound by nonionic, hydrophobic interactions can be eluted by reagents such as urea; and proteins that interact with lipid components can be eluted by nonionic detergents or organic solvents. For example, as urea is gradually added to preparations of purified adenovirus particles, they break down in an ordered, stepwise fashion which releases subvirus protein assemblies, revealing the composition of the particles. In the case of TMV, similar studies of capsid organization have been performed by renaturation of the capsid protein under various conditions (Figure 1.5). In addition to revealing structure,

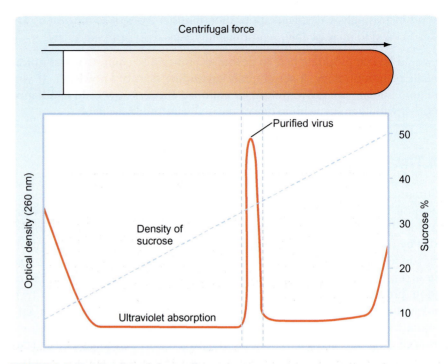

FIGURE 1.4 **Centrifugation of virus particles.**
A number of different sedimentation techniques can be used to study viruses. In rate-zonal centrifugation (shown here), virus particles are applied to the top of a preformed density gradient, that is, sucrose or a salt solution of increasing density from the top to the bottom of the tube. After a period of time in an ultracentrifuge, the gradient is separated into a number of fractions, which are analyzed for the presence of virus particles. The nucleic acid of the virus genome can be detected by its absorption of ultraviolet light. This method can be used both to purify virus particles or nucleic acids or to determine their sedimentation characteristics. In equilibrium or isopycnic centrifugation, the sample is present in a homologous mixture containing a dense salt such as cesium chloride. A density gradient forms in the tube during centrifugation, and the sample forms a band at a position in the tube equivalent to its own density. This method can be used to determine the density of virus particles and is sometimes used to purify plasmid DNA.

progressive denaturation can also be used to observe alteration or loss of antigenic sites on the surface of particles, and in this way a picture of the physical state of the particle can be developed. Proteins exposed on the surface of viruses can be labeled with various compounds (e.g., iodine) to indicate which parts of the protein are exposed and which are protected inside the particle or by lipid membranes. Cross-linking reagents are used to determine the spatial relationship of proteins and nucleic acids in intact viruses.

The physical properties of viruses can be determined by spectroscopy, using either ultraviolet light to examine the nucleic acid content of the particle or

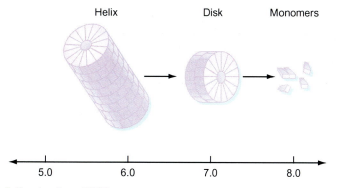

Helix   Disk   Monomers

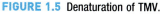

5.0   6.0   7.0   8.0

FIGURE 1.5 Denaturation of TMV.
The structure and stability of virus particles can be examined by progressive denaturation or renaturation studies. At any particular ionic strength, the purified capsid protein of TMV spontaneously assembles into different structures, dependent on the pH of the solution. At a pH of around 6.0, the particles formed have a helical structure very similar to infectious virus particles. As the pH is increased to about 7.0, disk-like structures are formed. At even higher pH values, individual capsid monomers fail to assemble into more complex structures.

visible light to determine its light-scattering properties. Electrophoresis of intact virus particles has yielded some limited information, but electrophoretic analysis of individual **virion** proteins by gel electrophoresis, and particularly of nucleic acid **genomes** (Chapter 3), has been far more valuable. However, the most important method for the investigation of virus structures has been the use of X-ray diffraction by crystals of purified viruses. This technique permits determination of the structure of virions at an atomic level. The complete structures of many viruses, representative of many of the major groups, have now been determined (Chapter 2).

Crystallography has improved our understanding of the function of virus particles considerably. However, a number of viruses are resistant to this type of investigation, a fact that highlights some of the problems inherent in this otherwise powerful technique. One problem is that the virus must first be purified to a high degree, otherwise specific information on the virus cannot be gathered. This presupposes that adequate quantities of the virus can be propagated in culture or obtained from infected tissues or patients and that a method is available to purify virus particles without loss of structural integrity. In a number of important cases, this requirement rules out further study (e.g., hepatitis C virus). The purified virus must also be able to form paracrystalline arrays large enough to cause significant diffraction of the radiation source. For some viruses, this is relatively straightforward, and crystals big enough to see with the naked eye and which diffract strongly are easily formed. This is particularly true for a number of plant viruses, such as TMV (which was first crystallized by Wendell Stanley in 1935) and turnip yellow mosaic virus (TYMV), the structures of which were among the first to be determined during the 1950s. It is significant that these two viruses represent the two fundamental types of virus particle: **helical** in the case of TMV and **icosahedral** for TYMV (see Chapter 2). In many cases, only microscopic crystals can be prepared. A partial answer to this problem is to use ever more powerful radiation sources that allow good data to be collected from small crystals. Powerful synchotron sources that generate intense beams of radiation have been built during the last few decades and are now used extensively for this purpose. However, there is a limit beyond which this brute force approach fails to yield further benefit. A number of important viruses refuse to crystallize; this is a particularly common problem with irregularly shaped viruses—for example, those which have an outer lipid **envelope**. Modifications of the basic diffraction technique (such as electron scattering by membrane-associated protein arrays) have helped to provide more information. One further limitation is that some of the largest virus particles, such as poxviruses, contain hundreds of different proteins and are at present too complex to be analyzed using these techniques.

Nuclear magnetic resonance (NMR) has been widely used to determine the atomic structure of all kinds of molecules, including proteins and nucleic acids. The limitation of this method is that only relatively small molecules can be analyzed before the signals obtained become so confusing that they are impossible to decipher with current technology. At present, the upper size limit for this technique restricts its use to molecules with a molecular weight of less than about 50,000, considerably less than even the smallest virus particles. Nevertheless, this method may well prove to be of value in the future, certainly for examining isolated virus proteins if not for intact **virions**.

Since the 1930s, electron microscopes have overcome the fundamental limitation of light microscopes: the inability to see individual virus particles owing to physical constraints caused by the wavelength of visible light illumination and the optics of the instruments. The first electron micrograph of a virus (TMV) was published in 1939. Over subsequent years, techniques were developed that allowed the direct examination of viruses at magnifications of over 100,000 times. The two fundamental types of electron microscope are the transmission electron microscope (TEM) and the scanning electron microscope (SEM) (Figure 1.6). In the late 1950s, Sydney Brenner and Robert Horne (among others) developed sophisticated techniques that enabled them to use electron microscopy to reveal many of the fine details of the structure of virus particles. One of the most valuable proved to be the use of electron-dense materials such as phosphotungstic acid or uranyl acetate to examine virus particles by negative staining. The small metal ions in the solution are able to penetrate the minute crevices between the protein

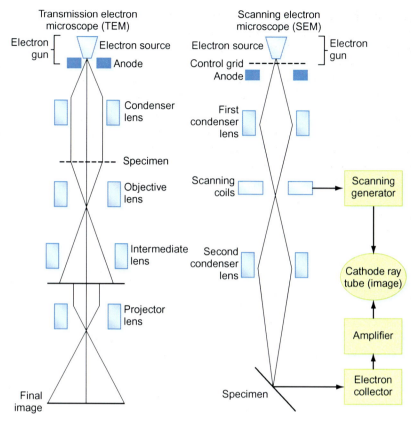

FIGURE 1.6 Electron microscopy.
This figure shows the working principles of transmission and scanning microscopes.

subunits in a virus **capsid** to reveal the fine structure of the particle. Using such data, Francis Crick and James Watson (1956) were the first to suggest that virus capsids are composed of numerous identical protein subunits arranged either in helical or cubic (**icosahedral**) symmetry. In 1962, Donald Caspar and Aaron Klug extended these observations and elucidated the fundamental principles of symmetry, which allow repeated protomers to form virus capsids, based on the principle of **quasi-equivalence** (see Chapter 2). This combined theoretical and practical approach has resulted in our current understanding of the structure of virus particles.

Although beautiful images with the appearance of three dimensions are produced by the SEM, for practical investigations of virus structure the higher magnifications achievable with the TEM have proved to be most valuable. Cryo-electron microscopy uses extremely low temperatures achieved by immersing the specimen in liquid nitrogen or helium to be able to increase the amount of radiation falling on the specimen and hence the resolution, without disrupting the structure. Cryo-electron tomography allows a computerized three-dimensional reconstruction of a sample from a series of two dimensional images taken at cryogenic temperatures. Conventional electron microscopy can resolve structures down to about 5 nm in size. Using these newer techniques it is possible to resolve structures of 2.5 nm; for comparison, a typical atomic diameter is 0.25 nm, a protein alpha-helix 1 nm, and a DNA double helix 2 nm. The most recent development is atomic force microscopy, with a resolution of less than 1 nm, which has been used to provide new insight into the structure of some viruses.

Two fundamental types of information can be obtained by electron microscopy of viruses: the absolute number of virus particles present in any preparation (total count) and the appearance and structure of the **virions** (see below). Electron microscopy can provide a rapid method of virus detection and diagnosis, but may give misleading information. Many cellular components (e.g., ribosomes) can resemble "virus-like particles," particularly in crude preparations. This difficulty can be overcome by using antisera specific for particular virus antigens conjugated to electron-dense markers such as the iron-containing protein ferritin or colloidal gold suspensions. This highly specific technique, known as immunoelectron microscopy, is useful in some cases as a rapid method for diagnosis.

MOLECULAR BIOLOGY

"Molecular biology" refers to a set of experimental techniques used to study the structure and function of biomolecules and their interactions. These techniques for manipulating nucleic acids and proteins *in vitro* (i.e., outside living cells or organisms) are not a new discipline but grew out of earlier developments in biochemistry and cell biology.

When the structure of a protein has been determined by X-ray crystallography or NMR (see above), the shape can be accurately modeled and explored in three dimensions on computers (Figure 1.7). Other powerful techniques for examining proteins have also been developed, notably mass spectrometry, which separates particles of different mass to charge ratios. The technique behind mass spectrometry dates back about a hundred years, but in the past few decades it has become a key technology for peptide sequencing, through which the identity of a protein can be deduced, either by sequencing overlapping peptides from a protein digest or by comparing a peptide sequence to known protein sequences stored in online databases so that the identity of the protein from which the peptide comes can be determined.

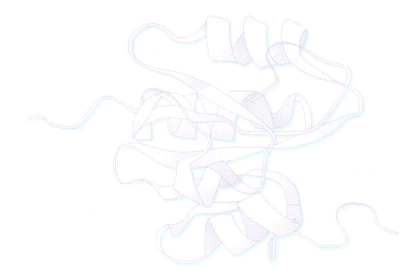

FIGURE 1.7 Three-dimensional structure of the DNA-binding domain of SV40 T-antigen. This image was reconstructed from NMR data using a computer.

Although protein chemistry has advanced greatly, the biggest impact from new technologies has come via nucleic acid–centered approaches, which have revolutionized virology and, to a large extent, shifted the focus of attention toward the virus genome, the nucleic acid comprising the entire genetic information of an organism. Initially, molecular biology techniques focused on direct analysis of nucleic acids using gel electrophoresis. The molecules examined were extracted directly from virus particles, a limiting procedure because of low sensitivity and the requirement for relatively large amounts of virus to be available. Molecular cloning (as opposed to biological cloning of pure strains of virus in culture) removed this limitation by enabling the production of essentially unlimited quantities of cloned virus nucleic acids, but was often difficult and laborious to achieve. By increasing the sensitivity of nucleic acid detection using methods such as Southern blotting for DNA, northern blotting for RNA (and the equivalent western blotting for proteins) (Figure 1.8), further advances were made, but another big step was the development of the polymerase chain reaction (PCR) in the mid-1980s (Figure 1.9). PCR allows detection (and cloning if required, e.g., for protein production) of tiny quantities of nucleic acid from virus particles, or from complex mixtures found in virus-infected cells. This technique has continued to be developed, for example, using reverse-transcription (RT-PCR) to copy RNA into DNA which can be amplified, and more recently by methods such as digital PCR (dPCR), an amazingly sensitive technique used to measure tiny quantities in nucleic acid in clinical samples. Perhaps the biggest advance of all was the combination of PCR with nucleotide sequencing methods.

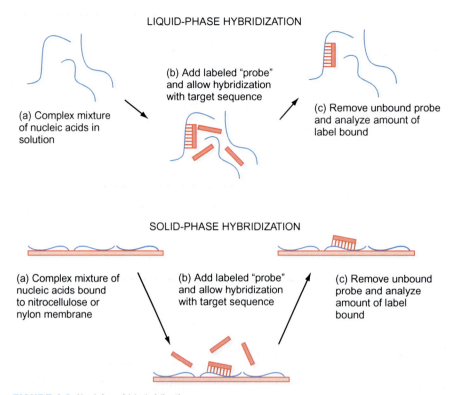

FIGURE 1.8 Nucleic acid hybridization.

Nucleic acid hybridization relies on the specificity of base-pairing which allows a labeled nucleic acid probe to pick out a complementary target sequence from a complex mixture of sequences in the test sample. The label used to identify the probe may be a radioisotope or a nonisotopic label such as an enzyme or photochemical. Hybridization may be performed with both the probe and test sequences in the liquid phase (top) or with the test sequences bound to a solid phase, usually a nitrocellulose or nylon membrane (below). Both methods may be used to quantify the amount of the test sequence present, but solid-phase hybridization is also used to locate the position of sequences immobilized on the membrane. Plaque and colony hybridization are used to locate recombinant molecules directly from a mixture of bacterial colonies or **bacteriophage** plaques on an agar plate. Northern and Southern blotting are used to detect RNA and DNA, respectively, after transfer of these molecules from gels following separation by electrophoresis (c.f. western blotting, Figure 1.2).

"Genomics" is the study of the composition and function of the genetic material of organisms. Virus genomics began with the first complete sequence of a virus genome (bacteriophage ØX174 in 1977). Initially, sequencing nucleic acids was a manual operation and was hard work. Gradually, the technology improved and this led to the evolution of what are sometimes called "next generation sequencing" methods. These include the ability to sequence both short RNA molecules such as mRNAs and intact

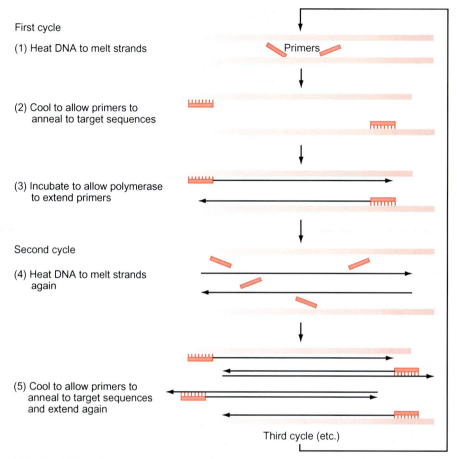

First cycle

(1) Heat DNA to melt strands

Primers

(2) Cool to allow primers to anneal to target sequences

(3) Incubate to allow polymerase to extend primers

Second cycle

(4) Heat DNA to melt strands again

(5) Cool to allow primers to anneal to target sequences and extend again

Third cycle (etc.)

FIGURE 1.9 Polymerase chain reaction (PCR).
The PCR relies on the specificity of base-pairing between short synthetic oligonucleotide probes and complementary sequences in a complex mixture of nucleic acids to prime DNA synthesis using a thermostable DNA polymerase. Multiple cycles of primer annealing, extension, and thermal denaturation are carried out in an automated process, resulting in a massive amplification of the target sequence located between the two primers ($2n$-fold increase after n cycles of amplification, i.e., over a million copies after 20 cycles).

RNA virus genomes via "RNAseq"—using RT-PCR to convert RNA into a DNA copy as the first step. Other current technologies include:

- Pyrosequencing (also known as 454 sequencing after the original robotic machine used for the process), which greatly reduced not only the time taken to sequence a genome but also the cost.
- Ion torrent sequencing, based on the detection of hydrogen ions that are released during the polymerization of DNA.

Table 1.1 Genomic Comparison of Different Organisms

Organism	Genome Size	Number of Genes
Porcine circovirus	1,759 nt	2
Hepatitis B virus	3,200 bp	4
SV40	5,200 bp	6
Herpes simplex virus	152 kbp (152,000 bp)	77
Pandoravirus	2.4 Mbp (2,473,870 bp)	2,500
Escherichia coli	4.6 Mbp (4,600,000 bp)	3,200
Yeast	12.1 Mbp (12,100,000 bp)	6,000
Caenorhabditis elegans	97 Mbp (97,000,000 bp)	19,000
Arabidopsis	100 Mbp (100,000,000 bp)	25,000
Drosophila	137 Mbp (137,000,000 bp)	13,000
Mouse	2,600 Mbp	25,000
Human	3,200 Mbp	25,000

■ Illumina dye sequencing, another highly parallel method which again uses optical detection of the chemical reactions occurring as DNA is copied in a highly automated process.

In terms of virology, it is not important to understand the details of how all these different methods work, but it is vital to appreciate the impact that they have had (and continue to have) on the subject. Vast international databases of nucleotide and protein sequence information have now been compiled, and these can be rapidly accessed by computers to compare newly determined sequences with those whose function may have been studied in great detail. Now, the complete genome sequences of thousands of different viruses have been published, with more appearing almost weekly (Table 1.1). But there is a problem. These methods have produced enormous quantities of data—many times larger than any biologist has ever had to deal with before. This has required the development of new parallel fields such as bioinformatics—a broad term first used in 1979 to describe the application of statistics and computer science to molecular biology—which can imply anything from artificial intelligence and robotics to genome analysis. As well as simply storing and displaying information, by spotting patterns, bioinformatics can infer function from simple digital information and so is central to all areas of modern biology (Figure 1.10). The current trend is to lump related information together and to analyze it in a holistic way rather than one molecule at a time. This is referred to as "omics"—approaches such as genomics or proteomics combine many of the above methods to examine all of the genes (genome) or proteins (proteome) in a given organism or environment, for example, the Vaccinia virus genome (the nucleotide sequence and all the genes in Vaccinia virus) and the Mimivirus proteome (all the

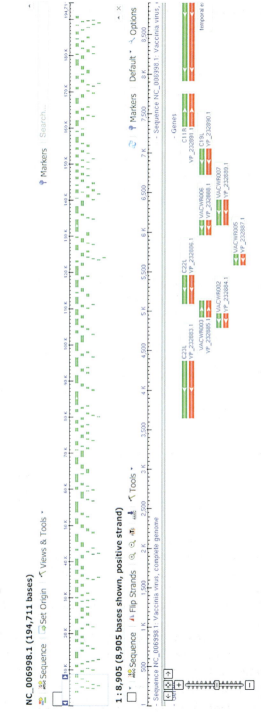

FIGURE 1.10 Bioinformatics.

Computers are used to store and process digitized information from nucleic acid sequences. This figure shows an analysis of all of the open reading frames (ORFs) present in Vaccinia virus. This complex data is freely available online in an interactive form via the NCBI database (tinyurl.com/ye8v9p3).

possible proteins produced by Mimivirus). Metagenomics enables the direct study of all the uncultured genomes from complex environments (e.g., the ocean or the human body), without requiring each one to be identified and cloned for fully sequenced. As well as looking at the complex mixture of viruses found in natural environments such as water or soil, work is currently underway to study the human virome—all of the genes from all of the viruses which infect humans. The new field of virus ecology has emerged in recent years and aims to analyze whole viral communities even though most of the viral sequences detected show no significant homology to previously known sequences.

There is one more important step that needs to be taken care. The digital information stored in nucleic acids is not of itself meaningful unless the functions of the molecules and their interactions can be understood. Functional genomics attempts to go beyond the description or cataloging of genes and proteins to understanding the interactions between all of the components of a biological system. This also includes methods such as co-localization studies, two-hybrid systems to study interactions between proteins, microarrays—two dimensional grids of probes on a solid substrate such as a glass slide to allow high throughput screening, and gene knockout experiments through mutagenesis or RNAi.

We have emerged from this history with a profound understanding of viruses and how intimately related they are to ourselves. However, the current pace of research in virology tells us that there is still far more that we need to know. The rest of this book will explain in detail what we already understand about viruses.

Further Reading

Alberts, B., Bray, D., Hopkin, K., Johnson, A., Lewis, J., Raff, M., et al., Essential Cell Biology. fourth ed. Garland Science, New York, ISBN 1317806271.

American Academy of Microbiology, 2013. Viruses throughout life & time: friends, foes, change agents. Available from: <http://academy.asm.org/index.php/browse-all-reports/5180-viruses-throughout-life-time-friends-foes-change-agents>.

Forterre, P., 2010. Defining life: the virus viewpoint. Orig. Life Evol. Biosph. 40 (2), 151−160. Available from: http://dx.doi.org/10.1007/s11084-010-9194-1.

Guerrero-Ferreira, R.C., Wright, E.R., 2013. Cryo-electron tomography of bacterial viruses. Virology 435 (1), 179−186.

Hatfull, G.F., Hendrix, R.W., 2011. Bacteriophages and their genomes. Curr. Opin. Virol. 1 (4), 298−303.

Lesk, A., 2014. Introduction to Bioinformatics. fourth ed. OUP, Oxford, ISBN: 0199651566.

Moreira, D., López-García, P., 2009. Ten reasons to exclude viruses from the tree of life. Nat. Rev. Microbiol. 7 (4), 306−311. Available from: http://dx.doi.org/10.1038/nrmicro2108.

PLoS Collections: The Human Microbiome Project Collection, 2012, Available from: <www.ploscollections.org/hmp>.

Raoult, D., Forterre, P., 2008. Redefining viruses: lessons from Mimivirus. Nat. Rev. Microbiol. 6 (4), 315−319. Available from: http://dx.doi.org/10.1038/nrmicro1858.

Villarreal, L., 2004. Are viruses alive? Sci. Am. 291 (6), 100−105.

Wooley, J.C., Godzik, A., Friedberg, I., 2010. A primer on metagenomics. PLoS Comput. Biol. 6 (2), e1000667. Available from: http://dx.doi.org/10.1371/journal.pcbi.1000667.

CHAPTER 2

Particles

Intended Learning Outcomes

On completing this chapter you should be able to:

- Explain the need for viruses to form outer coats.
- Discuss the role of symmetry in the formation of virus particles.
- Describe examples of different types of virus particles, from simple to more complex forms.

THE FUNCTION AND FORMATION OF VIRUS PARTICLES

Figure 2.1 shows an illustration of the approximate shapes and sizes of different families of viruses. Virus particles may range in size by nearly 100-fold, from around 17 nm (Porcine circovirus) to 1,200 nm in length (*Pithovirus sibericum*). The protein subunits in a virus **capsid** are redundant, that is, there are many copies in each particle. Damage to one or more capsid subunits may make that particular subunit nonfunctional, but rarely does limited damage destroy the infectivity of the entire particle. This helps make the capsid an effective barrier. The protein shells surrounding virus particles are very tough, about as strong as a hard plastic such as Perspex or Plexiglas, although they are only a billionth of a meter or so in diameter. They are also elastic and are able to deform by up to a third without breaking. This combination of strength, flexibility, and small size means that it is physically difficult (although not impossible) to break open virus particles by physical pressure.

The outer surface of the virus is also responsible for recognition of and interaction with the host cell. Initially, this takes the form of binding of a specific **virus-attachment protein** to a cellular **receptor** molecule. However, the capsid also has a role to play in initiating infection by delivering the **genome** in a form in which it can interact with the host cell. In some cases, this is a simple process that consists only of dumping the genome into the cytoplasm of the cell. In other cases, this process is much more complex, for example,

Principles of Molecular Virology. DOI: http://dx.doi.org/10.1016/B978-0-12-801946-7.00002-X

BOX 2.1 WHY BOTHER?

Why do viruses bother to form a particle to contain their genome? Some unusual and infectious agents such as viroids (see Chapter 8) don't. The fact that viruses pay the genetic and biochemical cost of encoding and assembling the components of a particle shows there must be some benefits. At the simplest level, the function of the outer shells of a virus particle is to protect the fragile nucleic acid genome from physical, chemical, or enzyme damage. After leaving the host cell, the virus enters a hostile environment that would quickly inactivate an unprotected genome. Nucleic acids are susceptible to physical damage such as shearing by mechanical forces, and to chemical modification by ultraviolet light (sunlight). The natural environment is full of nucleases from dead cells or deliberately secreted as defense against infection. In viruses with single-stranded genomes, breaking a single phosphodiester bond in the backbone of the genome or chemical modification of one nucleotide is sufficient to inactivate that virus particle, making replication of the genome impossible. Particles offer protection, but they also allow the virus to communicate with a host cell.

retroviruses carry out extensive modifications to the virus genome while it is still inside the particle, converting two molecules of single-stranded RNA to one molecule of double-stranded DNA before delivering it to the cell nucleus. Beyond protecting the genome, this function of the capsid is vital in allowing viruses to establish an infection.

In order to form particles, viruses must overcome two fundamental problems. First, they must assemble the particle using only the information available from the components that make up the particle itself. Second, virus particles form regular geometric shapes, even though the proteins from which they are made are irregular. How do these simple organisms solve these difficulties? The solutions to both problems lie in the rules of symmetry.

CAPSID SYMMETRY AND VIRUS ARCHITECTURE

It is possible to imagine a virus particle, the outer shell of which (the **capsid**) consists of a single, hollow protein molecule, which folds up trapping the virus **genome** inside. In practice, this arrangement cannot occur, for the following reason. The triplet nature of the genetic code means that three nucleotides (or base pairs, in the case of viruses with double-stranded genomes) are necessary to encode one amino acid. As parasites, viruses cannot use an alternative, more economical, genetic code because this could not be read by the host cell. Because the approximate molecular weight of a nucleotide triplet is 1,000 g/mol and the average molecular weight of a single amino acid is 150 g/mol, a nucleic acid can only encode a protein that is at most 15% of its own weight. For this reason, virus capsids must be made up of multiple protein molecules (subunit construction), and viruses must solve the problem of how these subunits are arranged to form a stable structure.

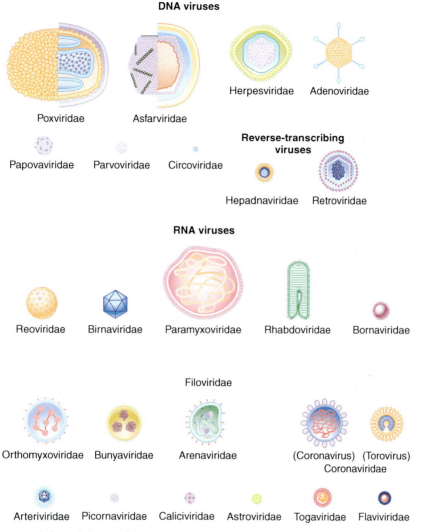

DNA viruses

Poxviridae   Asfarviridae   Herpesviridae   Adenoviridae

Papovaviridae   Parvoviridae   Circoviridae

Reverse-transcribing viruses

Hepadnaviridae   Retroviridae

RNA viruses

Reoviridae   Birnaviridae   Paramyxoviridae   Rhabdoviridae   Bornaviridae

Filoviridae

Orthomyxoviridae   Bunyaviridae   Arenaviridae   (Coronavirus)   (Torovirus)
Coronaviridae

Arteriviridae   Picornaviridae   Caliciviridae   Astroviridae   Togaviridae   Flaviviridae

FIGURE 2.1 Shapes and sizes of virus particles.
A diagram illustrating the shapes and sizes of viruses of families. The virions are drawn approximately to scale, but artistic license has been used in representing their structure. In some, the cross-sectional structures of capsid and envelope are shown, with a representation of the genome. For the very small virions, only their size and symmetry are depicted. *From F.A. Murphy, School of Veterinary Medicine, University of California, Davis. http://www.vetmed.ucdavis.edu/viruses/VirusDiagram.html*

In 1957, Fraenkel-Conrat and Williams showed that when mixtures of puri-fied tobacco mosaic virus (TMV) RNA and coat protein are incubated together, virus particles formed. The discovery that virus particles could form spontaneously from purified subunits without any extra information indicates that the particle is in the free energy minimum state and is the most energetically favored structure of the components. This inherent stability is an important feature of virus particles. Although some viruses are very fragile and unable to survive outside the protected host cell environment, many are able to persist for long periods, in some cases for years.

The forces that drive the assembly of virus particles include hydrophobic and electrostatic interactions. Only rarely are covalent bonds involved in holding together the subunits. In biological terms, this means that protein−protein, protein−nucleic acid, and protein−lipid interactions are involved. We now have a good understanding of general principles and repeated structural motifs that appear to govern the construction of many diverse, unrelated viruses. These are discussed below under the two main classes of virus structures: helical and icosahedral symmetry.

HELICAL CAPSIDS

TMV is representative of one of the two major structural classes seen in virus particles, those with helical symmetry. The simplest way to arrange multiple, identical protein subunits is to use rotational symmetry and to arrange the irregularly shaped proteins around the circumference of a circle to form a disk. Multiple disks can then be stacked on top of one another to form a cylinder, with the virus genome coated by the protein shell or contained in the hollow center of the cylinder. Denaturation studies of TMV suggest that this is the form this virus particle takes (see Chapter 1).

Closer examination of the TMV particle by X-ray crystallography reveals that the structure of the capsid actually consists of a helix rather than a pile of stacked disks. A helix can be defined mathematically by two parameters: its amplitude (diameter) and pitch (the distance covered by each complete turn of the helix) (Figure 2.2). Helices are simple structures formed by stacking repeated components with a constant association (amplitude and pitch) to one another. If this simple relationship is broken, a spiral forms rather than a helix, and a spiral is unsuitable for containing and protecting a virus genome. In terms of individual protein subunits, helices are described by the number of subunits per turn of the helix, μ, and the axial rise per subunit, p; therefore, the pitch of the helix, P, is equal to:

$$P = \mu \times p$$

For TMV, $\mu = 16.3$, that is there are 16.3 coat protein molecules per helix turn, and $p = 0.14$ nm. Therefore, the pitch of the TMV helix is $16.3 \times 0.14 = 2.28$ nm. TMV particles are rigid, rod-like structures, but some helical viruses demonstrate

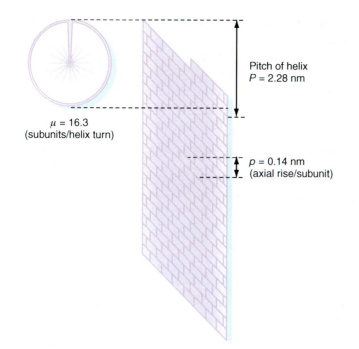

Pitch of helix
$P = 2.28$ nm

$\mu = 16.3$
(subunits/helix turn)

$p = 0.14$ nm
(axial rise/subunit)

FIGURE 2.2
TMV has a capsid consisting of many molecules of a single-coat protein arranged in a constant relationship, forming a helix with a pitch of 2.28 Å.

considerable flexibility, and longer helical virus particles are often curved or bent. Flexibility is an important property. Long helical particles are likely to be subject to shear forces and the ability to bend reduces the likelihood of breakage or damage.

Helical symmetry is a very useful way of arranging a single protein subunit to form a particle. This is confirmed by the large number of different types of virus that have evolved with this capsid structure. Among the simplest helical capsids are those of the bacteriophages of the family *Inoviridae*, such as M13 and fd. These phages are about 900 nm long and 9 nm in diameter, and their particles contain five proteins (Figure 2.3). The major coat protein is the product of phage gene 8 (g8p) and there are 2,700–3,000 copies of this protein per particle, together with approximately five copies each of four minor capsid proteins (g3p, g6p, g7p, and g9p) located at the ends of the filamentous particle. The primary structure of the major coat protein g8p explains many of the properties of these particles. Mature molecules of g8p consist of approximately 50 amino acid residues (a signal sequence of 23 amino acids is cleaved from the precursor protein during its translocation into the outer membrane of the host bacterium) and are almost entirely α-helical in

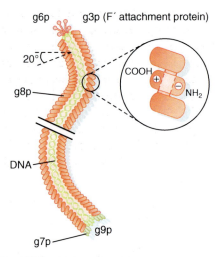

g6p   g3p (F′ attachment protein)

20°

COOH
+   −
NH₂

g8p

DNA

g9p

g7p

FIGURE 2.3 **Bacteriophage M13.**
Schematic representation of the bacteriophage M13 particle (*Inoviridae*). Major coat protein g8p is arranged helically, the subunits overlapping like the scales of a fish. Other capsid proteins required for the biological activity of the virion are located at either end of the particle. Inset shows the hydrophobic interactions between the g8p monomers (shaded region).

structure so that the molecule forms a short rod. There are three distinct regions within this rod. A negatively charged region at the amino-terminus that contains acidic amino acids forms the outer, hydrophilic surface of the virus particle, and a basic, positively charged region at the carboxy-terminus lines the inside of the protein cylinder adjacent to the negatively charged DNA genome. Between these two regions is a hydrophobic region that is responsible for interactions between the g8p subunits that allow the formation of and stabilize the phage particle (Figure 2.3). Inovirus particles are held together by these hydrophobic interactions between the coat protein subunits, demonstrated by the fact that the particles fall apart in the presence of chloroform, even though they do not contain a lipid component. The g8p subunits in successive turns of the helix interlock with the subunits in the turn below and are tilted at an angle of approximately 20° to the long axis of the particle, overlapping one another like the scales of a fish. The value of μ (protein subunits per complete helix turn) is 4.5, and p (axial rise per subunit) = 1.5 nm.

Because the phage DNA is packaged inside the core of the helical particle, the length of the particle is dependent on the length of the genome. In all inovirus preparations, *polyphage* (containing more than one genome length of DNA), *miniphage* (deleted forms containing 0.2−0.5 phage genome length of DNA), and *maxiphage* (genetically defective forms but containing more than one phage genome length of DNA) occur. The variable length of these filamentous particles has been exploited by molecular biologists to develop the M13 genome as a cloning vector. Insertion of foreign DNA into the genome results in recombinant phage particles that are longer than the wild-type filaments. Unlike most viruses, there is no sharp cut-off genome length at which the genome can no longer be packaged into the particle, however as the M13 genome size increases, the efficiency of replication declines. While recombinant phage genomes 1−10% longer than the wild-type do not appear to be significantly disadvantaged, those 10−50% longer than the wild-type replicate significantly more slowly. Above a 50% increase over the normal genome length it becomes progressively more difficult to isolate recombinant phage.

The structure of the inovirus capsid also explains the events that occur on infection of suitable bacterial host cells. Inovirus phages are male-specific, that is, they require the F pilus on the surface of *Escherichia coli* for infection. The first event in infection is an interaction between g3p located at one end of the filament together with g6p and the end of the F pilus. This interaction causes a conformational change in g8p. Initially, its structure changes from 100% α-helix to 85% α-helix, causing the filament to shorten. The end of the particle attached to the F pilus flares open, exposing the phage DNA. Subsequently, a second conformational change in the g8p subunits reduces its α-helical content from 85% to 50%, causing the phage particle to form a hollow spheroid about 40 nm in diameter and expelling the phage DNA, initiating the infection of the host cell.

Many plant viruses show helical symmetry (Appendix 2). These particles vary from approximately 100 nm (Tobravirus) to approximately 1,000 nm (Closterovirus) in length. The best studied example is, as stated above, TMV from the Tobamovirus group. Quite why so many groups of plant virus have evolved this structure is not clear, but it may be related either to the biology of the host plant cell or alternatively to the way in which they are transmitted between hosts.

Unlike plant viruses, helical, nonenveloped animal viruses do not exist. A large number of animal viruses are based on helical symmetry, but they all have an outer lipid envelope. The reason for this is again probably due to host cell biology and virus transmission mechanisms. There are too many viruses with this structure to list individually, but this category includes many of the best known human pathogens, such as influenza virus (*Orthomyxoviridae*), mumps and measles viruses (*Paramyxoviridae*), and rabies virus (*Rhabdoviridae*). All possess single-stranded, negative-sense RNA genomes (see Chapter 3). The molecular design of all of these viruses is similar. The virus nucleic acid and a basic, nucleic-acid-binding protein interact in infected cells to form a helical nucleocapsid. This protein−RNA complex protects the fragile virus genome from physical and chemical damage, but also provides vital functions associated with virus replication. The envelope and its associated proteins are derived from the membranes of the host cell and are added to the nucleocapsid core of the virus during replication (see Chapter 4).

Some of these helical, **enveloped** animal viruses are relatively simple in structure—for example, rabies virus and the closely related vesicular stomatitis virus (VSV) (Figure 2.4). These viruses are built up around the negative-sense RNA **genome**, which in rhabdoviruses is about 11,000 nucleotides (11 kilobases [kb]) long. The RNA genome and basic nucleocapsid (N) protein interact to form a helical structure with a pitch of approximately 5 nm, which, together with two nonstructural proteins, L and NS (which form the

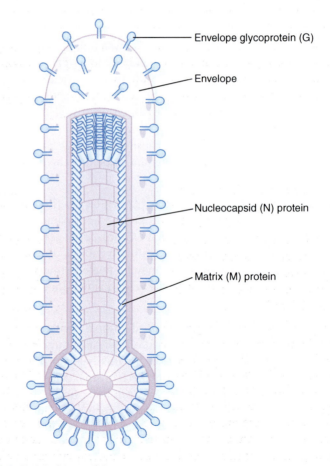

Envelope glycoprotein (G)

Envelope

Nucleocapsid (N) protein

Matrix (M) protein

FIGURE 2.4 Rhabdovirus particle.
Rhabdovirus particles, such as those of VSV, have an inner helical nucleocapsid surrounded by an outer lipid envelope and its associated glycoproteins.

virus polymerase; see Chapter 4), makes up the core of the virus particle. There are 30–35 turns of the nucleoprotein helix in the core, which is about 180 nm long and 80 nm in diameter. The individual N protein monomers are approximately $9 \times 5 \times 3$ nm, and each covers about nine nucleotides of the RNA genome. As in the case of the filamentous phage particles described above, the role of the N protein is to stabilize the RNA genome and protect it from chemical, physical, and enzyme damage. In common with most enveloped viruses, the nucleocapsid is surrounded by an amorphous layer with no visible structure that interacts with both the core and the overlying lipid envelope linking them together. This is known as the matrix. The matrix (M) protein is usually the most abundant protein in the virus particle; for example, there are approximately 1,800 copies of the M protein, 1,250 copies of the N protein, and 400 G protein trimers in VSV particles. The lipid envelope and its associated proteins are discussed in more detail later.

Many different groups of viruses have evolved with helical symmetry. Simple viruses with small genomes use this architecture to provide protection for the genome without the need to encode multiple different capsid proteins. More complex virus particles utilize this structure as the basis of the virus particle but elaborate on it with additional layers of protein and lipid.

ICOSAHEDRAL (ISOMETRIC) CAPSIDS

An alternative way of building a virus capsid is to arrange protein subunits in the form of a hollow quasi-spherical ("resembling a sphere") structure, enclosing the genome within it. The rules for arranging subunits on the surface of a solid are a little more complex than those for building a helix. In theory, a number of solid shapes can be constructed from repeated subunits—for example, a tetrahedron (four triangular faces), a cube (six square faces), an octahedron (eight triangular faces), a dodecahedron (12 pentagonal faces), and an icosahedron, a solid shape consisting of 20 triangular faces arranged around the surface of a sphere (Figure 2.5). Early in the 1960s, direct examination of a number of small "spherical" viruses by electron microscopy revealed that they appeared to have icosahedral symmetry. At first sight, it is not obvious why this pattern should have been "chosen" by diverse virus groups; however, although in theory it is possible to construct virus capsids based on simpler symmetrical arrangements, such as tetrahedra or cubes, there are practical reasons why this does not occur. As described above, it is more economic in terms of genetics to design a capsid based on a large number of identical, repeated protein subunits rather than fewer, larger subunits. It is unlikely that a simple tetrahedron consisting of four identical protein molecules would be large enough to contain even the smallest virus genome. If it were, it is probable that the gaps between the

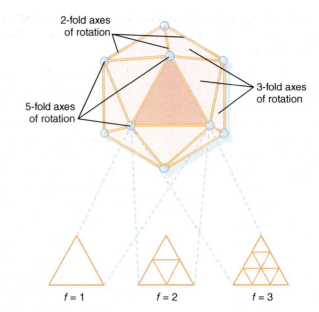

2-fold axes
of rotation

3-fold axes
of rotation

5-fold axes
of rotation

$f = 1$ $f = 2$ $f = 3$

FIGURE 2.5 Icosahedral symmetry
Illustration of the 2−3−5 symmetry of an icosahedron. More complex (higher order) icosahedra can be
defined by the triangulation number of the structure, $T = f^2 \times P$. Regular icosahedra have faces
consisting of equilateral triangles and are formed when the value of P is 1 or 3. All other values of P
give rise to more complex structures with either a left-hand or right-hand skew.

subunits would be so large that the particle would be leaky and fail to carry
out its primary function of protecting the virus genome.

In order to construct a capsid from repeated subunits, a virus must "know"
the rules that dictate how these are arranged. For an **icosahedron**, the rules
are based on the rotational symmetry of the solid, known as 2−3−5 symmetry, which has the following features (Figure 2.5):

- An axis of two-fold rotational symmetry through the center of each edge.
- An axis of three-fold rotational symmetry through the center of each face.
- An axis of five-fold rotational symmetry through the center of each
  corner (vertex).

Because protein molecules are irregularly shaped and are not regular equilateral triangles, the simplest icosahedral capsids are built up by using three
identical subunits to form each triangular face. This means that 60 identical
subunits are required to form a complete capsid (3 subunits per face, 20
faces). A few simple virus particles are constructed in this way; for example,
bacteriophages of the family *Microviridae*, such as φX174. An empty precursor particle called the **procapsid** is formed during assembly of this

bacteriophage. Assembly of the procapsid requires the presence of the two scaffolding proteins which are structural components of the procapsid but are not found in the mature virion.

In most cases, analysis reveals that icosahedral virus capsids contain more than 60 subunits, for the reasons of genetic economy given above. This causes a problem. A regular **icosahedron** composed of 60 identical subunits is a very stable structure because all the subunits are equivalently bonded (i.e., they show the same spacing relative to one another and each occupies the minimum free energy state). With more than 60 subunits it is impossible for them all to be arranged completely symmetrically with exactly equivalent bonds to all their neighbors, as a true regular icosahedron consists of only 20 subunits. To solve this problem, in 1962 Caspar and Klug proposed the idea of **quasi-equivalence**. Their idea was that subunits in nearly the same local environment form nearly equivalent bonds with their neighbors, permitting self-assembly of icosahedral capsids from multiple subunits. In the case of these higher order icosahedra, the symmetry of the particle is defined by the **triangulation number** of the icosahedron (Figure 2.5). The triangulation number, T, is defined by:

$$T = f^2 \times P$$

where f is the number of subdivisions of each side of the triangular face, and f^2 is the number of subtriangles on each face; $P = h^2 + hk + k^2$, where h and k are any distinct, nonnegative integers. This means that values of T fall into the series 1, 3, 4, 7, 9, 12, 13, 16, 19, 21, 25, 27, 28, and so on. When $P = 1$ or 3, a regular icosahedron is formed. All other values of P give rise to icosahedra of the "skew" class, where the subtriangles making up the icosahedron are not symmetrically arranged with respect to the edge of each face (Figure 2.6).

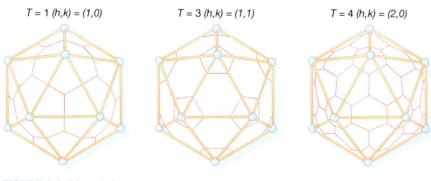

FIGURE 2.6 Triangulation numbers.
Icosahedra with triangulation numbers of 1, 3, and 4.

Detailed structures of icosahedral virus particles with $T = 1$ (*Microviridae*, e.g., ϕX174), $T = 3$ (many insect, plant, and animal RNA viruses), $T = 4$ (*Togaviridae*), and $T = 7$ (the heads of the tailed **bacteriophages** such as λ) have all been determined. With larger more complex viruses there is uncertainly—the triangulation number of large and complex Mimivirus particles could have any one of nine values between 972 and 1,200.

Virus particles with large triangulation numbers use different kinds of subunit assemblies for the faces and vertices of the icosahedron, and also use internal scaffolding proteins which act as a framework. These direct the assembly of the capsid, typically by bringing together preformed subassemblies of proteins (see discussion of ϕX174 above). Variations on the theme of icosahedral symmetry occur again and again in virus particles. For example, geminivirus particles consist of a fused pair of $T = 1$ icosahedra joined where one pentamer is absent from each icosahedron (hence their name, from the twins of Greek mythology, Castor and Pollux). Geminivirus particles consist of 110 capsid protein subunits and one molecule of ss(+)sense DNA of ~2.7 kb (Figure 2.7; Chapter 3). Elements of icosahedral symmetry occur frequently as part of larger assemblies of proteins (see "Complex Virus Structures").

The capsids of picornaviruses (*Picornaviridae*) provide a good illustration of the construction of icosahedral viruses. Detailed atomic structures of the capsids of a number of different picornaviruses have been determined. These include poliovirus, foot-and-mouth disease virus (FMDV), human rhinovirus, and several others. This work has revealed that the structure of these virus particles is remarkably similar to those of many other genetically unrelated

FIGURE 2.7 Geminivirus structure.
Geminivirus particles consist of "twinned" icosahedra, fused together at one of the pentameric vertices (corners). This allows some members of this family to contain a bipartite genome (see Chapter 3).

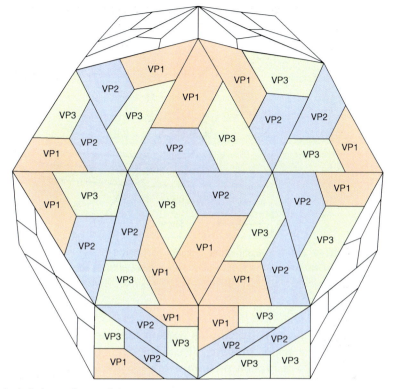

FIGURE 2.8 Icosahedral picornavirus particles.

Picornavirus particles are icosahedral structures with triangulation number $T = 3$. Three virus proteins (VP1, 2, and 3) comprise the surface of the particle. A fourth protein, VP4, is not exposed on the surface of the virion but is present in each of the 60 repeated units that make up the capsid.

viruses, such as insect viruses of the family *Nodaviridae* and plant viruses from the comovirus group. All these virus groups have icosahedral capsids approximately 30 nm diameter with **triangulation number** $T = 3$ (Figure 2.8). The capsid is composed of 60 repeated subassemblies of proteins, each containing three major subunits, VP1, VP2, and VP3. This means that there are $60 \times 3 = 180$ surface monomers in the entire picornavirus particle. All three proteins are based on a similar structure, consisting of 150–200 amino acid residues in what has been described as an "eight-strand antiparallel β-barrel" (Figure 2.9). This subunit structure has been found in all $T = 3$ icosahedral RNA virus capsids which have been examined so far (e.g., picornaviruses, comoviruses, nepoviruses), possibly showing distant evolutionary relationships between distinct virus families.

Knowledge of the structure of these $T = 3$ capsids also reveals information about the way in which they are assembled and the function of the mature

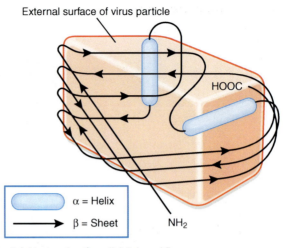

External surface of virus particle

HOOC

α = Helix

β = Sheet

NH₂

FIGURE 2.9 The "eight-strand antiparallel β-barrel."
The "eight-strand antiparallel β-barrel" subunit structure found in all $T = 3$ icosahedral RNA virus capsids.

capsid. Picornavirus capsids contain four structural proteins. In addition to the three major proteins VP1−3 (above), there is a small fourth protein, VP4. VP4 is located predominantly on the inside of the capsid and is not exposed at the surface of the particle. The way in which the four capsid proteins are processed from the initial **polyprotein** (see Chapter 5) was discovered by biochemical studies of picornavirus-infected cells (Figure 2.10). VP4 is formed from cleavage of the VP0 precursor into VP2 + VP4 late in assembly and is myristoylated at its amino-terminus (i.e., it is modified after translation by the covalent attachment of myristic acid, a 14-carbon unsaturated fatty acid). Five VP4 monomers form a hydrophobic micelle, driving the assembly of a pentameric subassembly. There is biochemical evidence that these pentamers, which form the vertices of the mature capsid, are a major precursor in the assembly of the particle; hence, the chemistry, structure, and symmetry of the proteins that make up the picornavirus capsid reveal how the assembly is driven.

Because they are the cause of a number of important human diseases, picornaviruses have been studied intensively by virologists. This interest has resulted in an outpouring of knowledge about these structurally simple viruses. Detailed knowledge of the structure and surface geometry of rhinoviruses has revealed lots about their interaction with host cells and with the immune system. In recent years, much has been learned not only about these viruses but also about the identity of their cellular **receptors** (see Chapter 4). The information from these experiments has been used to identify a number

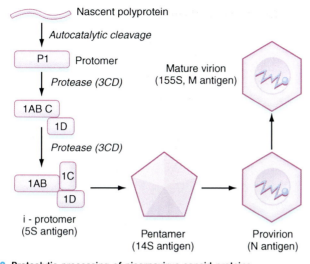

Nascent polyprotein

Autocatalytic cleavage

P1   Protomer

Protease (3CD)

1AB C
1D

Protease (3CD)

1AB   1C
1D

i - protomer
(5S antigen)

Mature virion
(155S, M antigen)

Pentamer
(14S antigen)

Provirion
(N antigen)

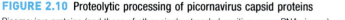

FIGURE 2.10 Proteolytic processing of picornavirus capsid proteins
Picornavirus proteins (and those of other single-stranded positive-sense RNA viruses) are produced by cleavage of a long polyprotein into the final products needed for replication and capsid formation (see Chapter 5).

of discrete antibody-neutralization sites on the surface of the virus particle. Some of these correspond to continuous linear regions of the primary amino acid sequence of the capsid proteins; others, known as conformational sites, result from separated stretches of amino acids coming together in the mature virus. With the elucidation of detailed picornavirus capsid structures, these regions have now been physically identified on the surface of the particle. They correspond primarily to hydrophilic, exposed loops of amino acid sequence, readily accessible to antibody binding and which are repeated on each of the pentameric subassemblies of the capsid. Now that the physical constraints on these sites are known, this type of information is being used to artificially manipulate them, even to build "antigenic chimeras" with the structural properties of one virus but expressing crucial antigenic sites from another.

ENVELOPED VIRUSES

So far, this chapter has concentrated on the structure of "naked" virus particles, that is, those in which the capsid proteins are exposed to the external environment. These viruses escape from infected cells at the end of the replication cycle when the cell dies, breaks down, and lyses, releasing the virions that have been built up internally. This simple strategy has drawbacks. In some circumstances, it is wasteful, resulting in the premature death of the

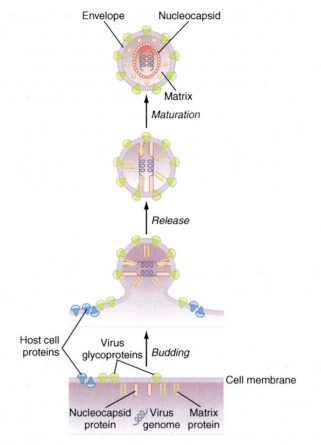

Envelope   Nucleocapsid

Matrix

Maturation

Release

Host cell   Virus   *Budding*
proteins   glycoproteins

Cell membrane

Nucleocapsid   Virus   Matrix
protein   genome   protein

FIGURE 2.11 Budding of enveloped particles.
Enveloped virus particles are formed by budding through a host cell membrane, during which the particle becomes coated with a lipid bilayer derived from the cell membrane. For some viruses, assembly of the structure of the particle and budding occur simultaneously, whereas in others a preformed core pushes out through the membrane.

cell and reducing the possibilities for persistent or latent infections. Many viruses have devised strategies to exit their host cell without causing its total destruction. The difficulty with this is that all living cells are covered by a membrane composed of a lipid bilayer. The viability of the cell depends on the integrity of this membrane. Viruses leaving the cell must, therefore, allow this membrane to remain intact. This is achieved by extrusion (**budding**) of the particle through the membrane, during which the particle becomes coated in a lipid **envelope** derived from the host cell membrane. The virus envelope is therefore similar in composition to the host cell membrane (Figure 2.11).

The particle structure underlying the envelope may be based on helical or icosahedral symmetry and may be formed before or as the virus leaves the cell. In the majority of cases, enveloped viruses use cellular membranes as sites allowing them to direct assembly. Formation of the particle inside the cell, maturation, and release are in many cases a continuous process. The site of assembly varies for different viruses. Not all use the cell surface membrane; many use cytoplasmic membranes such as the Golgi apparatus. Others, such as herpesviruses, which replicate in the nucleus, may utilize the nuclear membrane. In these cases, the virus is usually extruded into some sort of vacuole, in which it is transported to the cell surface and subsequently released. These points are discussed in more detail in Chapter 4.

If the virus particle became covered in a smooth, unbroken lipid bilayer, this would be its downfall. Such a coating is effectively inert, and, although effective as a protective layer preventing desiccation of or enzyme damage to the particle, it would not permit recognition of receptor molecules on the host cell. So viruses modify their lipid envelopes with several classes of proteins which are associated in one of three ways with the envelope (Figure 2.12). These can be summarized as follows:

- Matrix proteins. These are internal virion proteins whose function is to link the internal nucleocapsid assembly to the envelope. Matrix proteins are not usually glycosylated and are often very abundant, for example, in retroviruses they comprise approximately 30% of the total weight of the virion. Some matrix proteins contain transmembrane anchor domains. Others are associated with the membrane by hydrophobic patches on their surface or by protein−protein interactions with envelope glycoproteins.
- Glycoproteins. These proteins are anchored to the envelope and can be subdivided into two further types:
  - *External glycoproteins* are anchored in the envelope by a single transmembrane domain, or alternatively by interacting with a transmembrane protein. Most of the structure of the protein is on the outside of the membrane, sometimes with a short internal tail. Often, individual glycoprotein monomers associate with each other to form the multimeric "spikes" visible in electron micrographs on the surface of many enveloped viruses. These proteins are usually modified by glycosylation, which may be either N- or O-linked, and many are heavily glycosylated—up to 75% of the protein by weight may consist of sugars added posttranslationally. External glycoproteins are usually the major antigens of enveloped viruses and provide contact with the external environment, serving a number of important functions, for example, influenza virus

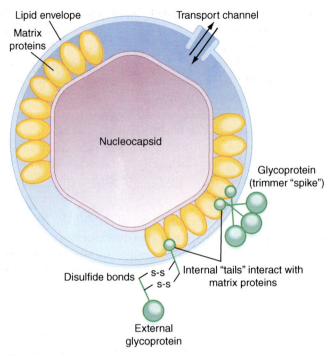

Lipid envelope

Matrix proteins

Transport channel

Nucleocapsid

Glycoprotein (trimmer "spike")

Disulfide bonds

S-S
S-S

Internal "tails" interact with matrix proteins

External glycoprotein

FIGURE 2.12 Envelope proteins.
Several classes of proteins are associated with virus envelopes. Matrix proteins link the envelope to the core of the particle. Virus-encoded glycoproteins inserted into the envelope serve several functions. External glycoproteins are responsible for receptor recognition and binding, while transmembrane proteins act as transport channels across the envelope. Host-cell-derived proteins are also sometimes found to be associated with the envelope, usually in small amounts.

    hemagglutinin is required for receptor binding, membrane **fusion**, and **hemagglutination**.

- *Transmembrane proteins* contain multiple hydrophobic domains that may form a protein-lined channel through the envelope. This enables the virus to control the permeability of the membrane (e.g., ion channels). Such proteins are often important in modifying the internal environment of the **virion**, permitting or even driving biochemical changes necessary for **maturation** of the particle and development of infectivity (e.g., the influenza virus M2 protein).

While there are many enveloped vertebrate viruses, only a few plant viruses have lipid envelopes. Most of these belong to the *Rhabdoviridae* family, whose structure has already been discussed (see "Helical Capsids"). Except for plant rhabdoviruses, only a few bunyaviruses which infect plants and members of the *Tospovirus* genus have outer lipid envelopes. The low number of enveloped plant viruses probably reflects aspects of host cell biology, in particular the mechanism of release of the virus from the infected cell, which requires a breach in the rigid cell wall. This limitation does not apply to viruses of prokaryotes, where there are a number of enveloped virus families (e.g., the *Cystoviridae*, *Fuselloviridae*, *Lipothrixviridae*, and *Plasmaviridae*).

COMPLEX VIRUS STRUCTURES

BOX 2.2 WHAT ARE YOU LOOKING AT?

When learning about virus structures, a picture certainly paints a thousand words (although it is important to be able to describe structures accurately in words too). There are some great resources freely available online which illustrate virus structures in detail, going far beyond what is possible in a single textbook. The problem is that such sites change regularly, so the best strategy is a simple web search for "virus structure" which will bring you the best and most up-to-date resources as these are created. Top Tip: "virus structure images" and "virus structure video" searches work pretty well too!

The majority of viruses can be fitted into one of the three structural groups outlined above (i.e., helical symmetry, icosahedral symmetry, or enveloped viruses). However, there are many viruses whose structure is more complex. In these cases, although the general principles of symmetry already described are often used to build parts of the virus shell (this term being appropriate here because such viruses often consist of several layers of protein and lipid), the larger and more complex viruses cannot be simply defined by a mathematical equation as can a simple helix or icosahedron. Because of the complexity of some of these viruses, they have defied attempts to determine detailed atomic structures using the techniques described in Chapter 1.

An example of such a group is the *Poxviridae*. These viruses have oval or "brick-shaped" particles 200–400 nm long. These particles are so large that they were first seen using high-resolution optical microscopes in 1886 and were thought at that time to be bacterial spores. The external surface of the virion is ridged in parallel rows, sometimes arranged helically. The particles are extremely complex and contain more than 100 different proteins (Figure 2.13). During replication, two forms of particles are produced, extracellular forms that contain two membranes and intracellular particles that have only an inner membrane. Poxviruses and other viruses with complex structures (such as African swine fever virus) obtain their membranes in a different way from "simple" enveloped viruses such as retroviruses or influenza. Rather than budding at the cell surface or into an intracellular compartment, thus acquiring a single membrane, these complex viruses are wrapped

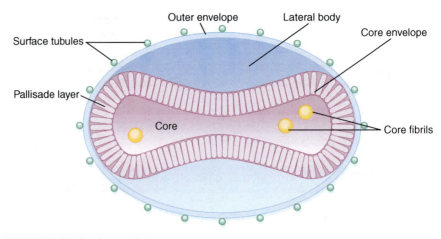

FIGURE 2.13 Poxvirus particle.
Poxvirus particles are the most complex virions known and contain more than 100 virus-encoded proteins, arranged in a variety of internal and external structures.

by the endoplasmic reticulum, thus acquiring two layers of membrane. Under the electron microscope, sections of poxvirus particle reveal an outer surface of the virion composed of lipid and protein. This layer surrounds the core, which is biconcave (dumbbell-shaped), and two "lateral bodies" whose function is unknown. The core is composed of a tightly compressed nucleo-protein, and the double-stranded DNA genome is wound around it. Antigenically, poxviruses are very complex, inducing both specific and cross-reacting antibodies, hence the possibility of vaccinating against one disease with another virus (e.g., the use of vaccinia virus to immunize against small-pox [variola] virus). Poxviruses and a number of other complex viruses also emphasize the true complexity of some viruses—there are at least 10 enzymes present in poxvirus particles, mostly involved in nucleic acid metab-olism/genome replication.

Poxviruses are among the most complex particles known. They are at one end of the scale of complexity and are included here as a counterbalance to the descriptions of the simpler viruses given above. In between these extremes lie intermediate examples of such as the tailed bacteriophages. The order *Caudovirales* (see Chapter 3), consisting of the families *Myoviridae*, *Siphoviridae*, and *Podoviridae*, has been extensively studied for several reasons. These viruses are easy to propagate in bacterial cells, can be obtained in high titers and are easily purified, making biochemical and structural studies com-paratively straightforward. The head of these particles consists of an icosahe-dral shell with $T = 7$ symmetry attached by a collar to a contractile, helical tail. At the end of the tail is a plate that functions in attachment to the

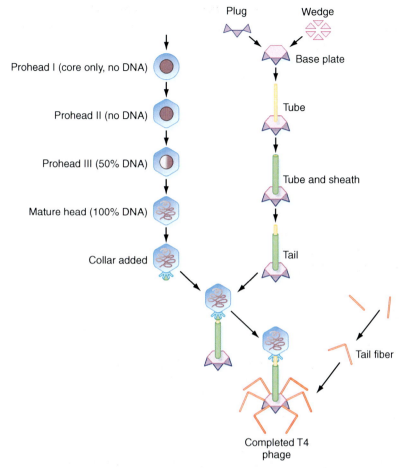

FIGURE 2.14 Assembly pathway of bacteriophage T4.
Simplified version of the assembly of bacteriophage T4 particles (*Myoviridae*). The head and tail sections are assembled separately and are brought together at a relatively late stage. This complex process was painstakingly worked out by the isolation of phage mutants in each of the virus genes involved. In addition to the major structural proteins, a number of minor "scaffolding" proteins are involved in guiding the formation of the complex particle.

bacterial host and also in **penetration** of the bacterial host cell wall by means of lysozyme-like enzymes associated with the plate. In addition, thin protein fibers attached to the tail plate are involved in binding to **receptor** molecules in the wall of the host cell. Inside this compound structure there are also internal proteins and polyamines associated with the genomic DNA in the head, and an internal tube structure inside the outer sheath of the helical tail. The sections of the particle are put together by separate assembly pathways for the head and tail sections inside infected cells, and these come together at a late stage to make up the infectious **virion** (Figure 2.14). These viruses illustrate how complex particles can be built up from the simple principles outlined above. Another example of this phenomenon is provided by the structure of geminivirus particles, which consist of two, twinned $T = 1$ icosahedra. Each icosahedron has one morphological subunit missing, and the icosahedra are joined at the point such that the mature particle contains 110 protein monomers arranged in 22 morphological subunits.

Baculoviruses have attracted interest for a number of reasons. They are natural pathogens of arthropods, and naturally occurring as well as genetically manipulated baculoviruses are under investigation as biological control agents for insect pests. In addition, occluded baculoviruses are used as expression vectors to produce large amounts of recombinant proteins. These complex viruses contain 12−30 structural proteins and consist of a rod-like (hence, "baculo") **nucleocapsid** that is 30−90 nm in diameter and 200−450 nm long, and contains the 90- to 230-kbp double-stranded DNA genome (Figure 2.15). The nucleocapsid is surrounded by an **envelope**, outside which there may be a crystalline protein matrix. If this outer protein

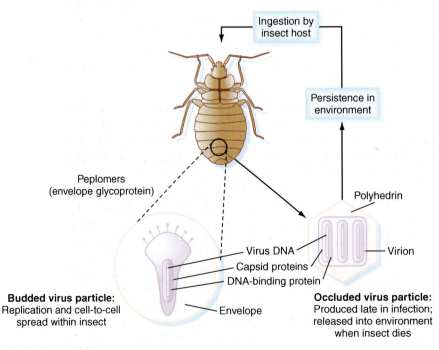

FIGURE 2.15 Baculovirus particles.
Some baculovirus particles exist in two forms: a relatively simple "budded" form found within the host insect and a crystalline, protein occluded form responsible for environmental persistence.

shell is present, the whole structure is referred to as an "occlusion body" and the virus is said to be occluded (Figure 2.15). There are two genera of occluded baculoviruses: the *Nucleopolyhedrovirus* genus, with polyhedral occlusions 1,000–15,000 nm in diameter and which may contain multiple nucleocapsids within the envelope (e.g., *Autographa californica nucleopolyhedrovirus*), and the *Granulovirus* genus, with ellipsoidal occlusions 200–500 nm diameter. The function of these large occlusion bodies is to allow the particle to resist adverse environmental conditions which enables the virus to persist in soil or on plant materials for extended periods of time while waiting to be ingested by a new host. These viruses can be regarded as being literally armor plated. Interestingly, the strategy of producing occluded particles appears to have evolved independently in at least three groups of insect viruses. In addition to the baculoviruses, occluded particles are also produced by insect reoviruses (cytoplasmic polyhedrosis viruses) and poxviruses (entomopoxviruses). However, this resistant coating would be the undoing of the virus if it were not removed at an appropriate time to allow replication to proceed. To solve this problem, the occlusion body is alkali-labile and dissolves in the high pH of the insect midgut, releasing the nucleocapsid and allowing it to infect the host. Although the structure of the entire particle has not been completely determined, it is known that the occlusion body is composed of many copies of a single protein of approximately 245 amino acids—polyhedrin. To form the occlusion body, this single gene product is hyperexpressed late in infection by a very strong transcriptional promoter. Cloned foreign genes can be expressed by the polyhedrin promoter. This has allowed baculoviruses to be engineered as expression vectors.

The final example of complex virus structure to be considered is Mimivirus (Figure 2.16). This is the largest virus presently known, about half of a micrometer (0.0005 mm) in diameter. The structure of the Mimivirus capsid has recently been determined by a combination of electron tomography and cryo-scanning electron microscopy—sophisticated image reconstruction techniques based on electron microscopy—and atomic force microscopy—which has a resolution of 1 nm (1×10^{-9} m) (see Chapter 1). Rather than standard icosahedral symmetry, Mimivirus has another configuration called five-fold symmetry. Like an icosahedron, the Mimivirus capsid also has 20 faces. However, unlike a regular icosahedron, five faces of the capsid are slightly different than the others and surround a special structure at one of the corners called the "stargate." In order to deliver the large 1.2 million base pair genome to the host cell, the stargate opens up and allows it to leave the capsid. The extreme resolution of atomic force microscopy has also revealed a regular pattern of small holes regularly spaced throughout the capsid. This unique feature among viruses and their function is unknown at present.

BOX 2.3 YES, I AM OBSESSED WITH MIMIVIRUS, AND HERE'S WHY

In Chapter 1, I described how Mimivirus blurs the boundaries between viruses and bacteria. The structure of Mimivirus particles is far more complex than that of any other virus studied so far. Not just in the size, triangulation number or complexity, but also in the way the particle functions, with the "stargate" at one corner opening up to allow the massive DNA genome to pass into the host cell. This is amazing science, and reading the suggested references at the end of this chapter will tell you not only the details of this monster, but also how these achievements were made using cutting edge scientific tools.

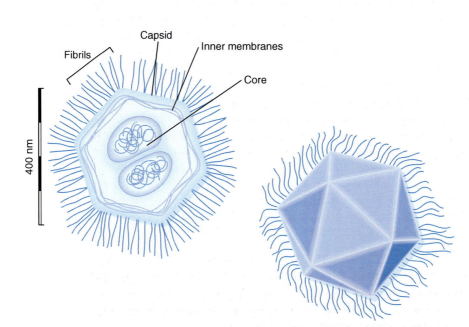

FIGURE 2.16 Mimivirus.
Although occluded baculoviruses might be bigger in diameter, Mimivirus is the most complex virus particle investigated in depth so far. One corner (vertex) of the particle is modified to form the "stargate" which allows the large DNA genome to pass into the host cell.

PROTEIN–NUCLEIC ACID INTERACTIONS AND GENOME PACKAGING

The primary function of the virus particle is to contain and protect the **genome** before delivering it to the appropriate host cell. To do this, the proteins of the **capsid** must interact with the nucleic acid genome. Once again the physical constraints of incorporating a relatively large nucleic acid molecule into a relatively small capsid present considerable problems which must be overcome. In most cases, the linear virus genome when stretched out in solution is at least an order of magnitude longer than the diameter of the capsid. Simply folding the genome in order to stuff it into such a confined space is quite a feat in itself but the problem is made worse by the fact that repulsion by the cumulative negative electrostatic charges on the phosphate groups of the nucleotide backbone means that the genome resists being crammed into a small space. Viruses overcome this difficulty by packaging, along with the genome, various positively charged molecules to counteract this negative charge repulsion. These include small, positively charged ions (Na, Mg, K, etc.), polyamines, and various nucleic-acid-binding proteins. Some of these proteins are virus-encoded and contain amino acids with basic side-chains, such as arginine and lysine, which interact with the genome. There are many examples of such proteins—for example, retrovirus NC and rhabdovirus N (nucleocapsid) proteins and influenza virus NP protein (nucleoprotein). Some viruses with double-stranded DNA genomes have basic histone-like molecules closely associated with the DNA. Again, some of these are virus-encoded (e.g., adenovirus polypeptide VII). In other cases, however, the virus may use cellular proteins, for example, the polyomavirus genome assumes a chromatin-like structure in association with four cellular histone proteins (H2A, H2B, H3, and H4), similar to that of the host cell genome.

The second problem the virus must overcome is how to achieve the specificity required to select and encapsidate (package) the virus genome from the background noise of cellular nucleic acids. In most cases, by the late stages of virus infection when assembly of virus particles occurs (see Chapter 4), transcription of cellular genes has been reduced and a large pool of virus genomes has accumulated. Overproduction of virus nucleic acids eases but does not eliminate the problem of specific genome packaging. A virus-encoded capsid or nucleocapsid protein is required to achieve specificity, and many viruses, even those with relatively short, compact genomes such as retroviruses and rhabdoviruses, encode this type of protein.

Viruses with segmented genomes (see Chapter 3) face further problems. Not only must they encapsidate only virus nucleic acid and exclude host cell molecules, but they must also attempt to package at least one of each of the required genome segments. During particle assembly, viruses frequently make mistakes. These can be measured by particle:infectivity ratios—the ratio of the total number of particles in a virus preparation (counted by electron microscopy) to the number of particles able to give rise to infectious progeny (measured by plaque or limiting dilution assays). This value is in some cases found to be several thousand particles to each infectious virion and only rarely approaches a ratio of 1:1. However, calculations show that viruses such as influenza have far lower particle:infectivity ratios than could be achieved by random packaging of eight different genome segments. Until recently, the process by which this was achieved was poorly understood, but a clearer picture has emerged of a mechanism for specifically packaging a full genome, mediated by *cis*-acting packaging signals in the vRNAs. Consequently, genome packaging, previously thought to be a random process in influenza virus, turns out to be quite highly ordered. Rotaviruses, with their 11 genome segments, face an even more complex problem, and it is not clear how they have solved this issue.

Although retrovirus genomes are not segmented, retrovirus particles are only infectious if they contain two complete copies of the genome (because of the need for two copies during reverse transcription). Different types of retrovirus solve the problem of how to package not one but two genomes in different ways.

On the other side of the packaging equation are the specific nucleotide sequences in the **genome** (the **packaging signal**) which permit the virus to select genomic nucleic acids from the cellular background. The packaging signal from a number of virus genomes has been identified. Examples are the ψ (psi) signal in retrovirus genomes which has been used to package synthetic "retrovirus vector" genomes into a virus particle, and the sequences responsible for packaging the genomes of several DNA virus genomes (e.g., some adenoviruses and herpesviruses) which have been clearly and unambiguously defined. However, it is clear from a number of different approaches that accurate and efficient genome packaging requires information not only from the linear nucleotide sequence of the genome but also from regions of secondary structure formed by the folding of the genomic nucleic acid into complex forms. In many cases, attempts to find a unique, linear packaging signal in virus genomes have failed. The probable reason for this is that the key to the specificity of genome packaging in most viruses lies in the secondary structure of the genome.

Like many other aspects of virus assembly, the way in which packaging is controlled is, in many cases, not well understood. However, the key must lie in the specific molecular interactions between the **genome** and the **capsid**. Until recently, the physical structure of virus genomes within virus particles has been poorly studied, although the genetics of packaging have been extensively investigated. This is a shame, because it is unlikely that we will be able fully to appreciate this important aspect of virus replication without this information, but it is understandable because the techniques used to determine the structure of virus capsids (e.g., X-ray diffraction) only rarely reveal any information about the state of the genome within its protein shell. However, in some cases detailed knowledge about the mechanism and specificity of genome encapsidation is now available. These include both viruses with helical symmetry and some with icosahedral symmetry.

Undoubtedly the best understood packaging mechanism is that of the (+) sense RNA helical plant virus, TMV. This is due to the relative simplicity of

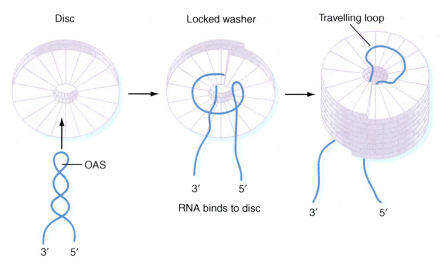

Disc   Locked washer   Travelling loop

OAS

3′   5′

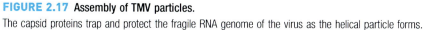

RNA binds to disc   3′   5′

3′ 5′

FIGURE 2.17 **Assembly of TMV particles.**
The capsid proteins trap and protect the fragile RNA genome of the virus as the helical particle forms.

this virus, which has only a single major coat protein and will spontaneously assemble from its purified RNA and protein components *in vitro*. In the case of TMV, particle assembly is initiated by association of preformed aggregates of coat protein molecules ("disks") with residues 5,444–5,518 in the 6.4 kb RNA genome, known as the origin of assembly sequence (OAS) (Figure 2.17). The flat disks have 17 subunits per ring, close to the 16.34 subunits per turn found in the mature virus particle. In fact, the disks are not completely symmetrical, as they have a pronounced polarity. Assembly begins when a disk interacts with the OAS in genomic RNA. This converts the disks to a helical "locked washer" structure, each of which contains 3′ coat protein subunits. Further disks add to this structure, switching to the "locked washer" conformation. RNA is drawn into the assembling structure in what is known as a "traveling loop," which gives the common name to this mechanism of particle formation. The vRNA is trapped and subsequently buried in the middle of the disk as the helix grows. Extension of the helical structure occurs in both directions but at unequal rates. Growth in the 5′ direction is rapid because a disk can add straight to the protein filament and the traveling loop of RNA is drawn up through it. Growth in the 3′ direction is slower because the RNA has to be threaded through the disk before it can add to the structure.

Bacteriophage M13 is another helical virus where protein–nucleic acid interactions in the virus particle are relatively simple to understand (Figure 2.3). The primary sequence of the g8p molecule determines the orientation of the protein in the capsid. In simple terms, the inner surface of the rod-like phage capsid is positively charged and interacts with the negatively charged genome, while the outer surface of the cylindrical capsid is negatively charged. However, the way in which the capsid protein and genome are brought together is a little more complex. During replication, the genomic DNA is associated with a nonstructural DNA-binding protein, g5p. This is the most abundant of all virus proteins in an M13-infected *E. coli* cell, and it coats the newly replicated single-stranded phage DNA, forming an intracellular rod-like structure similar to the mature phage particle but somewhat longer and thicker ($1{,}100 \times 16$ nm). The function of this protein is to protect the genome from host cell nucleases and to interrupt genome replication, sequestering newly formed strands as substrates for encapsidation. Newly synthesized coat protein monomers (g8p) are associated with the inner (cytoplasmic) membrane of the cell, and it is at this site that assembly of the virus particle occurs. The g5p coating is stripped off as the particle passes out through the membrane and is exchanged for the mature g8p coat (plus the accessory proteins). The forces that drive this process are not fully understood, but the protein–nucleic acid interactions that occur appear to be rather simple and involve opposing electrostatic charges and the stacking of the DNA bases between the planar side-chains of the proteins. This is confirmed by the variable length of the M13 genome and its ability to freely encapsidate extra genetic material.

Protein–nucleic acid interactions in other helical viruses, such as rhabdoviruses, are rather more complex. In most enveloped helical viruses, a nucleoprotein core forms first which is then coated by matrix proteins, the envelope, and its associated glycoproteins (Figure 2.4). The structure of the core shows cross-striations 4.5–5.0 nm apart, each of these presumably equating to one turn of the protein–RNA complex (rather like TMV). Assembly of rhabdoviruses follows a well-orchestrated program. It begins with RNA and the nucleocapsid (N) protein stretched out as a ribbon. The ribbon curls into a tight ring and then is physically forced to curl into larger rings that eventually form the helical trunk at the center of the particle. Matrix (M) protein subunits bind on the outside of the nucleocapsid, rigidifying the bullet-shaped tip and then the trunk of the particle, and create a triangularly packed platform for binding glycoprotein (G) trimers and the envelope membrane, all in a coherent operation during budding of the particle from the infected cell.

Rather less is known about the arrangement of the genome inside virus most particles with icosahedral symmetry. There are a few exceptions to this statement. These are the $T = 3$ icosahedral RNA viruses whose subunits consist largely of the "eight-strand antiparallel β-barrel" structural motif, discussed earlier. In these viruses, positively charged inward-projecting arms of the capsid proteins interact with the RNA in the center of the particle. In bean pod mottle virus (BPMV), a $T = 3$ comovirus with a bipartite genome (see Chapter 3), X-ray crystallography has shown that the RNA is folded in such a way that it assumes icosahedral symmetry, corresponding to that of the capsid surrounding it. The regions that contact the capsid proteins are single stranded and appear to interact by electrostatic forces rather than by covalent bonds. The atomic structure of ϕX174 also shows that a portion of the DNA genome interacts with arginine residues exposed on the inner surface of the capsid in a manner similar to BPMV.

A consensus about the physical state of nucleic acids within icosahedral virus capsids appears to be emerging. Just as the icosahedral capsids of many genetically unrelated viruses are based on monomers with a common "eight-strand antiparallel β-barrel" structural motif, the genomes inside also appear to display icosahedral symmetry, the vertices of which interact with basic amino acid residues on the inner surface of the capsid. These common structural motifs may explain how viruses selectively package the required genomic nucleic acids and may even offer opportunities to design specific drugs to inhibit these vital interactions.

VIRUS RECEPTORS: RECOGNITION AND BINDING

Cellular receptor molecules used by a number of different viruses from diverse groups have now been identified. The interaction of the outer surface of a virus with a cellular receptor is a major event in determining the subsequent events in replication and the outcome of infections. It is this binding event that activates inert extracellular virus particles and initiates the replication cycle. Receptor binding is considered in detail in Chapter 4.

OTHER INTERACTIONS OF THE VIRUS CAPSID WITH THE HOST CELL

As described earlier, the function of the virus capsid is not only to protect the genome but also to deliver it to a suitable host cell, and more specifically, to the appropriate compartment of the host cell (in the case of eukaryote hosts) in order to allow replication to proceed. One example is the nucleocapsid protein of viruses which replicate in the nucleus of the host cell. These molecules contain within their primary amino acid sequences "nuclear localization signals" that are responsible for the migration of the virus genome plus its associated proteins into the nucleus where replication can occur. Again, these events are discussed in Chapter 4.

Virions are not inert structures. Many virus particles contain one or more enzymatic activities, although in most cases these are not active outside the biochemical environment of the host cell. All viruses with negative-sense RNA genomes must carry with them a virus-specific, RNA-dependent RNA polymerase because most eukaryotic cells have no mechanism for RNA-dependent RNA polymerization so genome replication could not occur if this enzyme was not included in the virus particle. Reverse transcription of retrovirus genomes occurs inside a particulate complex and not free in solution. The more complex DNA viruses (e.g., herpesviruses and poxviruses) carry a multiplicity of enzymes, mostly concerned with some aspect of nucleic acid metabolism.

Our fundamental structural understanding of the structure and function of virus particles is beginning to have benefits in new ways. Some of these have been long anticipated, such as rational vaccine design based on structural data rather than trial and error, but others are new, such as the role this knowledge is playing in nanotechnology and the design of molecular machines.

SUMMARY

This chapter is not intended to be a complete list of all known virus structures, but it does try to illustrate the principles that control the assembly of viruses and the difficulties of studying these tiny objects. There are a number of repeated structural patterns found in many different virus groups. The most obvious is the division virus structures into those based on helical or icosahedral symmetry. More subtly, common protein structures such as the "eight-strand antiparallel β-barrel" structural motif found in many $T = 3$ icosahedral virus capsids and the icosahedral folded RNA genome present inside some of these viruses are beginning to emerge. Virus particles are not inert structures. Many are armed with a variety of enzymes that carry out a range of complex reactions, most frequently involved in the replication of the genome. Most importantly, virus particles are designed to interact with host cell receptors to initiate the process of infection which will be considered in detail in subsequent chapters.

Further Reading

Brown, J.C., Newcomb, W.W., Wertz, G.W., 2010. Helical virus structure: the case of the rhabdovirus bullet. Viruses 2, 995–1001.

Cherwa, J.E., Fane, B.A., 2009. Complete virion assembly with scaffolding proteins altered in the ability to perform a critical conformational switch. J. Virol. 83 (15), 7391–7396.

Hemminga, M.A., Vos, W.L., Nazarov, P.V., Koehorst, R.B., Wolfs, C.J., Spruijt, R.B., et al., 2010. Viruses: incredible nanomachines. New advances with filamentous phages. Eur. Biophys. J. 39 (4), 541–550.

Hutchinson, E.C., von Kirchbach, J.C., Gog, J.R., Digard, P., 2010. Genome packaging in influenza A virus. J. Gen. Virol. 91 (2), 313–328.

Ivanovska, I.L., De Pablo, P.J., Ibarra, B., Sgalari, G., MacKintosh, F.C., Carrascosa, J.L., et al., 2004. Bacteriophage capsids: tough nanoshells with complex elastic properties. Proc. Natl. Acad. Sci. USA 101, 7600–7605.

Johnson, S.F., Telesnitsky, A., 2010. Retroviral RNA dimerization and packaging: the what, how, when, where, and why. PLoS Pathog. 6 (10), e1001007.

Klug, A., 1999. The tobacco mosaic virus particle: structure and assembly. Philos. Trans. R. Soc. Lond. B Biol. Sci. 354 (1383), 531–535.

Mannige, R.V., Brooks, C.L., 2010. Periodic table of virus capsids: implications for natural selection and design. PLoS One 5 (3), e9423.

Maurer-Stroh, S., Eisenhaber, F., 2004. Myristoylation of viral and bacterial proteins. Trends Microbiol. 12, 178–185.

McDonald, S.M., Patton, J.T., 2011. Assortment and packaging of the segmented rotavirus genome. Trends Microbiol. 19 (3), 136–144.

McKenna, R., Xia, D., Willingmann, P., Ilag, L.L., Krishnaswamy, S., Rossmann, M.G., et al., 1992. Atomic structure of single-stranded DNA bacteriophage ϕX174 and its functional implications. Nature 355, 137–143.

Roberts, K.L., Smith, G.L., 2008. Vaccinia virus morphogenesis and dissemination. Trends Microbiol. 16 (10), 472–479.

Rohrmann, G.F., 2008. Baculovirus molecular biology. NCBI Bookshelf. <http://www.ncbi.nlm.nih.gov/bookshelf/br.fcgi?book=bacvir>.

Xiao, C., et al., 2009. Structural studies of the giant Mimivirus. PLoS Biol. 7 (4), e1000092.

CHAPTER 3

Genomes

Intended Learning Outcomes

On completing this chapter you should be able to:

- Describe the range of structures and compositions seen in virus genomes.
- Explain how the composition and structure of a genome affects the genetic mechanisms which operate on it.
- Discuss representative examples of viruses genomes to illustrate the range of genetic diversity seen in viruses.

THE STRUCTURE AND COMPLEXITY OF VIRUS GENOMES

Unlike the genomes of all cells, which are composed of DNA, virus genomes may contain their genetic information encoded in either DNA or RNA. The chemistry and structures of virus genomes are more varied than any of those seen in the entire bacterial, plant, or animal kingdoms. The nucleic acid making up the genome may be single stranded or double stranded, and it may have a linear, circular, or segmented structure. Single-stranded virus genomes may be either positive-sense (i.e., the same polarity or nucleotide sequence as the mRNA), negative-sense, or ambisense (a mixture of the two). Virus genomes range in size from approximately 2,500 nucleotides (nt) (e.g., the Geminivirus tobacco yellow dwarf virus at 2,580 nt of single-stranded DNA) to 2.8 million base pairs of double-stranded DNA in the case of Pandoravirus, which is nearly five times as large as the smallest bacterial genome (e.g., *Mycoplasma genitalum* at 580,000 bp). Some of the simpler bacteriophages are well-studied examples of the smallest and least complex genomes. At the other end of the scale, the genomes of the largest double-stranded DNA viruses such as herpesviruses and poxviruses are sufficiently complex to have escaped complete functional analysis yet, even though the

CONTENTS

59

Principles of Molecular Virology. DOI: http://dx.doi.org/10.1016/B978-0-12-801946-7.00003-1

complete nucleotide sequences of the genomes of a large number of examples have been known for many years.

Whatever the composition of a virus genome, they must follow one rule. Because viruses are obligate intracellular parasites only able to replicate inside the appropriate host cells, the genome must contain information encoded in a way that can be recognized and decoded by the particular type of host cell. The genetic code used by the virus must match or at least be recognized by the host organism. Similarly, the control signals that direct the expression of virus genes must be appropriate to the host. Many of the DNA viruses of **eukaryotes** closely resemble their host cells in terms of the biology of their genomes. Chapter 4 describes the ways in which virus genomes are replicated, and Chapter 5 deals in more detail with the mechanisms that regulate the expression of virus genetic information. The purpose of this chapter is to describe the diversity of virus genomes and to consider how and why this variation may have arisen.

Virus genome structures and nucleotide sequences have been intensively studied in recent decades because the power of recombinant DNA technology has focused much attention in this area. It would be wrong to present molecular biology as the only means of addressing unanswered problems in virology, but it would be equally foolish to ignore the opportunities that it offers and the explosion of knowledge that has resulted from it in recent years. As noted in Chapter 1, this has been matched by an explosion in bioinformatics techniques to process and make sense of all this data.

Some DNA virus genomes are complexed with cellular histones to form a **chromatin**-like structure inside the virus particle. Once inside the nucleus of the host cell, these genomes behave like miniature satellite chromosomes, controlled by cellular enzymes and the cell cycle:

- Vaccinia virus mRNAs were found to be polyadenylated at their 3′ ends by Kates in 1970—the first time this observation had been made in any organism.
- Split genes containing noncoding **introns**, protein-coding **exons**, and spliced mRNAs were first discovered in adenoviruses by Roberts and Sharp in 1977.

Introns in **prokaryotes** were first discovered in the genome of bacteriophage T4 in 1984. Several examples of this phenomenon have now been discovered in T4 and some other phages. This raises an important point. The conventional view is that prokaryote genomes are smaller and replicate faster than those of **eukaryotes** and hence can be regarded as "streamlined." The genome of phage T4 consists of 160 kbp of double-stranded DNA and is highly compressed, for example, **promoters** and translation control sequences are nested within the coding regions of overlapping upstream genes. The presence of introns in bacteriophage genomes, which are under constant ruthless pressure to exclude "junk sequences," suggests that these genetic elements must have evolved mechanisms to escape or neutralize this pressure and to persist as parasites within parasites. All virus genomes experience pressure to minimize their size. Viruses with prokaryotic hosts must be able to replicate sufficiently quickly to keep up with their host cells, and this is reflected in the compact nature of many (but not all) bacteriophages. Overlapping genes are common, and the maximum genetic capacity is compressed into the minimum genome size. In viruses with eukaryotic hosts there is also pressure on genome size. Here, however, the pressure is mainly from the packaging size of the virus particle (i.e., the amount of nucleic acid that can be incorporated into the virion). Therefore, these viruses commonly show highly compressed genetic information when compared with the low density of information in eukaryotic cellular genomes.

There are exceptions to this rule. Some bacteriophages (e.g., the family *Myoviridae*, such as T4) have relatively large genomes, up to 170 kbp. The genome of Mimivirus, at approximately 1.2 Mbp, contains around 1,200 open reading frames (ORFs), only 10% of which show any similarity to proteins of known function. Among viruses of eukaryotes, herpesviruses and poxviruses also have relatively large genomes, up to 235 kbp. It is notable that these virus genomes contain many genes involved in their own replication, particularly enzymes concerned with nucleic acid metabolism. These viruses partially escape the restrictions imposed by the biochemistry of the host cell by encoding additional biochemical equipment. The penalty is that they have to encode all the information necessary for a large and complex particle to package the genome—which is also an upward pressure on genome size. Later sections of this chapter contain detailed descriptions of both small, compact and large, complex virus genomes.

BOX 3.1 IT'S NOT THE SIZE OF YOUR GENOME THAT COUNTS, IT'S WHAT YOU DO WITH IT

Traditionally it was thought that virus genomes were smaller than bacterial genomes. Often that is true, but not always. So does having a bigger genome make a better virus? Not in my opinion. As discussed in this chapter, some virus genomes are as complex as bacterial genomes, and larger than some of the smaller ones. This means they have nearly the same capabilities as bacteria—but not quite. No virus genome contains all the genes needed to make ribosomes, so in the end they are still parasites. Personally, my admiration goes to those stripped down miniature marvels which contain only a handful of genes and yet still manage to take over a cell and replicate themselves successfully. Now that's impressive.

MOLECULAR GENETICS

As already described, the techniques of molecular biology have been a major influence on concentrating much attention on the virus genome. It is beyond the scope of this book to give detailed accounts of these methods. However, it is worth taking some time here to illustrate how some of these techniques have been applied to virology, remembering that these newer techniques are complementary to and do not replace the classical techniques of virology. Initially, any investigation of a virus genome will usually include questions about the following:

- Composition—DNA or RNA, single-stranded or double-stranded, linear or circular
- Size and number of segments
- Nucleotide sequence
- Terminal structures
- Coding capacity—ORFs
- Regulatory signals—transcription enhancers, promoters, and terminators

It is possible to separate the molecular analysis of virus genomes into two types of approach: physical analysis of structure and nucleotide sequence, essentially performed *in vitro*, and a more biological approach to examine the structure–function relationships of intact virus genomes and individual genetic elements, usually involving analysis of the virus phenotype *in vivo*.

The conventional starting point for the physical analysis of virus genomes has been the isolation of nucleic acids from virus preparations of varying degrees of purity. More recently the emphasis on extensive purification has declined as techniques of molecular cloning and amplification have become more advanced, allowing direct examination of genomes in complex mixtures of sequences without the need for laborious purification first. DNA virus genomes can be analyzed directly by restriction endonuclease digestion without resorting to molecular cloning, and this approach was achieved for the first time with SV40 DNA in 1971. The first pieces of DNA to be molecularly cloned were restriction fragments of bacteriophage λ DNA which were cloned into the DNA genome of SV40 by Berg and colleagues in 1972. This means that virus genomes were both the first cloning vectors and the first nucleic acids to be analyzed by these techniques. In 1977, the genome of bacteriophage ϕX174 was the first replicon to be completely sequenced.

Subsequently, phage genomes such as M13 were highly modified for use as vectors in DNA sequencing. The enzymology of RNA-specific nucleases was comparatively advanced at this time, such that a spectrum of enzymes with specific cleavage sites could be used to analyze and even determine the sequence of RNA virus genomes (the first short nucleotide sequences of tRNAs having been determined in the mid-1960s). However, direct analysis of RNA by these methods was laborious and notoriously difficult. RNA sequence analysis did not begin to advance rapidly until the widespread use of reverse transcriptase (isolated from retroviruses) to convert RNA into cDNA in the 1970s. Since the 1980s, polymerase chain reaction (PCR) has further accelerated the investigation of virus genomes (Chapter 1), and direct sequence analysis of RNA genomes via RT-PCR is now common.

In addition to molecular cloning, other techniques of molecular analysis have been historically valuable in virology. Direct analysis by electron microscopy, if calibrated with known standards, can be used to estimate the size of nucleic acid molecules. Hybridization of complementary nucleotide sequences can also be used in a number of ways to analyze virus genomes (Chapter 1). Perhaps the most important single technique has been gel electrophoresis (Figure 3.1). The earliest gel matrix employed for separating molecules was based on starch and gave relatively poor resolution. It is now most common

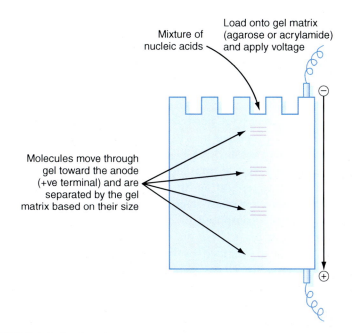

FIGURE 3.1 Gel electrophoresis.
In gel electrophoresis, a mixture of nucleic acids (or proteins) is applied to a gel, and they move through the gel matrix when an electric field is applied. The net negative charge due to the phosphate groups in the backbone of nucleic acid molecules results in their movement away from the cathode and toward the anode. Smaller molecules are able to slip though the gel matrix more easily and thus migrate farther than larger molecules, which are retarded, resulting in a net separation based on the size of the molecules.

to use agarose gels to separate large nucleic acid molecules, which may be very large indeed—several megabases (million base pairs) in the case of techniques such as pulsed-field gel electrophoresis, and polyacrylamide gel electrophoresis to separate smaller pieces (down to sizes of a few nucleotides). Apart from the fact that sequencing depends on the ability to separate molecules that differ from each other by only one nucleotide in length, gel electrophoresis has been of great value in analyzing intact virus genomes, particularly the analysis of viruses with segmented genomes. The most recent and most powerful sequence analysis techniques have done away with electrophoresis and rely on light detection from fluorescent compounds.

Phenotypic analysis of virus populations has long been a standard technique of virology. In modern terms, this might be considered as functional genomics. Examination of variant viruses and naturally occurring spontaneous mutants is an old method for determining the function of virus genes. Molecular biology has added to this the ability to design and create specific mutations, deletions, and recombinants *in vitro*. This site-directed mutagenesis is a very powerful tool. Although genetic coding capacity can be examined *in vitro* by the use of cell-free extracts to translate **mRNAs**, complete functional analysis of virus genomes can only be performed on intact viruses. Fortunately, the relative simplicity of most virus genomes (compared with even the simplest cell) offers a major advantage here—the ability to "rescue" infectious virus from purified or cloned nucleic acids. Infection of cells caused by nucleic acid alone is referred to as **transfection**.

Virus genomes that consist of positive-sense RNA are infectious when the purified RNA (vRNA) is applied to cells in the absence of any virus proteins. This is because positive-sense vRNA is essentially mRNA, and the first event in a normally infected cell is to translate the vRNA to make the virus proteins responsible for genome replication. In this case, direct introduction of RNA into cells circumvents the earliest stages of the replicative cycle (Chapter 4). Virus genomes that are composed of double-stranded DNA are also infectious. The events that occur here are a little more complex, as the virus genome must first be transcribed by host polymerases to produce mRNA. This is relatively simple for phage genomes introduced into **prokaryotes**, but for viruses that replicate in the nucleus of **eukaryotic** cells, such as herpesviruses, the DNA must first find its way to the appropriate cellular compartment. Most of the DNA that is introduced into cells by transfection is degraded by cellular nucleases. However, irrespective of its sequence, a small proportion of the newly introduced DNA finds its way into the nucleus, where it is transcribed by cellular polymerases.

Unexpectedly, cloned cDNA genomes of positive-sense RNA viruses (e.g., picornaviruses) are also infectious, although less efficient at infecting cells than the vRNA. This is presumably because the DNA is transcribed by cellular enzymes to make RNA. Synthetic RNA transcribed *in vitro* from the cDNA template of the genome is much more efficient at initiating infection. Such experiments are referred to as "reverse genetics"—that is, the manipulation of a virus via a cloned intermediate. Using these techniques, virus can be rescued from cloned genomes, including those that have been manipulated *in vitro*. Originally, this type of approach was not possible for analysis of viruses with negative-sense genomes. This is because all negative-sense virus particles contain a virus-specific polymerase. The first event when these virus genomes enter the cell is that the negative-sense genome is copied by the polymerase, forming either positive-sense transcripts that are used directly as **mRNA** or a double-stranded molecule, known either as the replicative intermediate or replicative form, which serves as a template for further rounds of mRNA synthesis. Therefore, because purified negative-sense genomes cannot be directly translated by the host cell and are not replicated in the absence of the virus polymerase, these genomes are inherently noninfectious. However, systems have now been developed that permit the rescue of viruses with negative-sense genomes from purified or cloned nucleic acids. All such systems rely on a ribonucleoprotein complex that can serve as a template for genome replication by RNA-dependent RNA polymerase, but they fall into one of two approaches:

- *In vitro* complex formation: Virus proteins purified from infected cells are mixed with RNA transcribed from cloned cDNAs to form complexes which are then introduced into susceptible cells to initiate an infection. This method has been used for paramyxoviruses, rhabdoviruses, and bunyaviruses.
- *In vivo* complex formation: Ribonucleoprotein complexes formed *in vitro* are introduced into cells infected with a helper virus strain. This method has been used for influenza virus, bunyaviruses, and double-stranded RNA viruses such as reoviruses and birnaviruses.

Such developments open up possibilities for genetic investigation of negative- and double-stranded RNA viruses that have not previously existed, and are of particular interest because of their potential for vaccine development (see Chapter 6).

VIRUS GENETICS

Although nucleotide sequencing now dominates the analysis of virus genomes, functional genetic analysis of animal viruses is based largely on the isolation and analysis of mutants, usually achieved using plaque purification ("biological cloning"). In the case of viruses for which no such systems exist (because they either are not cytopathic or do not replicate in culture), little genetic analysis was possible before the development of molecular genetics. However, certain tricks make it possible to extend standard genetic techniques to noncytopathic viruses:

- **Biochemical analysis:** Use of metabolic inhibitors to construct genetic "maps"; inhibitors of translation (such as puromycin and cycloheximide) and transcription (actinomycin D) can be used to decipher genetic regulatory mechanisms.
- **Focal immunoassays:** Replication of noncytopathic viruses visualized by immune staining to produce visual foci (e.g., human immunodeficiency virus, HIV).
- **Physical analysis:** Use of high-resolution electrophoresis to identify genetic polymorphisms of virus proteins or nucleic acids.
- **Transformed foci:** Production of transformed "foci" of cells by noncytopathic "focus-forming" viruses (e.g., DNA and RNA tumor viruses).

Various types of genetic maps can be constructed:

- **Recombination maps:** These represent an ordered sequence of mutations derived from the probability of recombination between two genetic markers, which is proportional to the distance between them— a classical genetic technique. This method works for viruses with nonsegmented genomes (DNA or RNA).
- **Reassortment maps (or groups):** In viruses with segmented genomes, the assignment of mutations to particular genome segments results in identification of genetically linked "reassortment groups" equivalent to individual genome segments.

VIRUS MUTANTS

"Mutant," "strain," "type," "variant," and even "isolate" are all terms used rather loosely by virologists to differentiate particular viruses from each other and from the original "parental," "wild-type," or "street" isolates of that virus. More accurately, these terms are generally applied as follows:

- **Strain:** Different lines or isolates of the same virus (e.g., from different geographical locations or patients).
- **Type:** Different serotypes of the same virus (e.g., various antibody neutralization phenotypes).
- **Variant:** A virus whose phenotype differs from the original wild-type strain but the genetic basis for the difference is not known, for example, a new clinical isolate from a patient.

Mutant viruses can arise spontaneously or be artificially constructed. In some viruses, mutation rates may be as high as 10^{-3} to 10^{-4} per incorporated nucleotide (e.g., in retroviruses such as HIV), whereas in others they may be as low as 10^{-8} to 10^{-11} (e.g., in herpesviruses), which is similar to the mutation rates seen in cellular DNA. These differences are due to the mechanism of genome replication, with error rates in RNA-dependent RNA polymerases generally being higher than DNA-dependent DNA polymerases. Some RNA virus polymerases do have proofreading functions, but in general mutation rates are higher in most RNA viruses than in DNA viruses. For a virus, mutations are a mixed blessing. The ability to generate antigenic variants that can escape the immune response is a clear advantage, but mutation also results in many defective particles, as most mutations are deleterious. In the most extreme cases (e.g., HIV), the error rate is 10^{-3} to 10^{-4} per nucleotide incorporated. The HIV genome is approximately 9.7 kb long; therefore, there will be 0.9−9.7 mutations in every genome copied. Hence, in this case, the wild-type virus actually consists of a fleeting majority type that dominates the dynamic equilibrium (i.e., the population of genomes) present in all cultures of the virus. These mixtures of molecular variants are known as quasispecies and also occur in other RNA viruses (e.g., picornaviruses). However, the majority of these variants will be noninfectious or seriously disadvantaged and are therefore rapidly weeded out of a replicating population. This mechanism is an important force in virus evolution (see "The Virome—Evolution and Epidemiology"). Historically, most genetic analysis of viruses has been performed on virus mutants isolated from mutagen-treated populations. Currently, defined and highly specific mutations in particular genes or control sequences (e.g., single nucleotide changes) are usually introduced using molecular methods such as PCR. This is known as site-directed or site-specific mutagenesis.

The disadvantage of determining nucleotide sequences as a primary tool for studying viruses is that the sequence alone frequently says little about the function of gene beyond what can be guessed from comparisons with similar genes in other organisms. To understand how the virus actually works functional analysis is required and this is where mutations, whether naturally occurring or artificially introduced, are valuable. The phenotype of a mutant virus depends on the type of mutation(s) it has and also upon the location of the mutation(s) within the genome. Each of the classes of mutations can occur naturally in viruses or may be artificially induced for experimental purposes:

- **Biochemical markers:** These include drug resistance mutations, mutations that result in altered virulence, polymorphisms resulting in altered electrophoretic mobility of proteins or nucleic acids, and altered sensitivity to inactivating agents.
- **Deletions:** Similar in some ways to nonsense mutants (below) but may include one or more virus genes and involve noncoding control regions of the genome (promoters, etc.). Spontaneous deletion mutants often

accumulate in virus populations as defective-interfering (DI) particles. These noninfectious but not necessarily genetically inert genomes are thought to be important in establishing the course and pathogenesis of certain virus infections (see Chapter 6). Genetic deletions can only revert to wild-type by recombination, which usually occurs at comparatively low frequencies. Deletion mutants are very useful for assigning structure–function relationships to virus genomes, as they are easily mapped by physical analysis.

- **Host range:** This term can refer either to whole animal hosts or to permissive cell types *in vitro*. Conditional mutants of this class have been isolated using amber-suppressor cells (mostly for phages but also for animal viruses using *in vitro* systems).

- **Nonsense:** These result from alteration of coding sequence of a protein to one of three translation stop codons (UAG, amber; UAA, ocher; UGA, opal). Translation is terminated, resulting in the production of an amino-terminal fragment of the protein. The phenotype of these mutations can be suppressed by propagation of virus in a cell (bacterial or, more recently, animal) with altered suppressor tRNAs. Nonsense mutations are rarely "leaky" (i.e., the normal function of the protein is completely obliterated) and can only revert to wild-type at the original site, so they usually have a low reversion frequency.

- **Plaque morphology:** Mutants may be either large-plaque mutants which replicate more rapidly than the wild-type, or small-plaque mutants, which are the opposite. Plaque size is often related to a temperature-sensitive (t.s.) phenotype. These mutants are often useful as unselected markers in multifactorial crosses.

- **Temperature-sensitive (t.s.):** This type of mutation is very useful as it allows the isolation of conditional-lethal mutations, a powerful means of examining virus genes that are essential for replication and whose function cannot otherwise be interrupted. Temperature-sensitive mutations usually result from mis-sense mutations in proteins (i.e., amino acid substitutions), resulting in proteins of full size with subtly altered conformation that can function at (permissive) low temperatures but not at (nonpermissive) higher ones. Generally, the mutant proteins are immunologically unaltered, which is frequently useful. These mutations are usually "leaky"—that is, some of the normal activity is retained even at nonpermissive temperatures. Protein function is often impaired, even at permissive temperatures, therefore a high frequency of reversion is often a problem with this type of mutation because the wild-type virus replicates faster than the mutant.

- **Cold-sensitive (c.s.):** These mutants are the opposite of temperature-sensitive mutants and are very useful in bacteriophages and plant viruses whose host cells can be propagated at low temperatures but are less useful for animal viruses because their host cells generally will not grow at significantly lower temperatures than normal.

- **Revertants:** Reverse mutation is a valid type of mutation in its own right. Most of the above classes can undergo reverse mutation, which may be either simple "back mutations" (i.e., correction of the original mutation) or second-site "compensatory mutations," which may be physically distant from the original mutation and not even necessarily in the same gene as the original mutation.

- **Suppression:** Suppression is the inhibition of a mutant phenotype by a second suppressor mutation, which may be either in the virus genome or in that of the host cell. This mechanism of suppression is not the same as the suppression of chain-terminating amber mutations by host-encoded suppressor tRNAs, which could be called "informational suppression." Genetic suppression results in an apparently wild-type phenotype from a virus which is still genetically mutant—a pseudorevertant. This phenomenon has been best studied in prokaryotic systems, but examples have been discovered in animal viruses—for example, reoviruses, vaccinia, and influenza—where suppression has been observed in an attenuated vaccine, leading to an apparently virulent virus. Suppression may also be important biologically in that it allows viruses to overcome the deleterious effects of mutations and therefore be positively selected.

Mutant viruses can appear to revert to their original phenotype by three pathways:

1. Back mutation of the original mutation to give a wild-type genotype/phenotype (true reversion).
2. A second, compensatory mutation that may occur in the same gene as the original mutation, thus correcting it—for example, a second frameshift mutation restoring the original reading frame (intragenic suppression).
3. A suppressor mutation in a different virus gene or a host gene (extragenic suppression).

GENETIC INTERACTIONS BETWEEN VIRUSES

Genetic interactions between viruses often occur naturally, as host organisms are frequently infected with more than one virus. These situations are generally too complicated to be analyzed successfully. Experimentally, genetic interactions can be analyzed by mixed infection (superinfection) of cells in culture. Two types of information can be obtained from such experiments:

- The assignment of mutants to functional groups known as complementation groups.
- The ordering of mutants into a linear genetic map by analysis of recombination frequencies.

Complementation results from the interaction of virus gene products during superinfection which results in production of one or both of the parental viruses being increased while both viruses remain unchanged genetically. In this situation, one of the viruses in a mixed infection provides a functional gene product for another virus which is defective for that function (Figure 3.2). If both mutants are defective in the same function, enhancement of replication does not occur and the two mutants are said to be in the same complementation group. The importance of this test is that it allows functional analysis of unknown mutations if the biochemical basis of any one of the mutations in a particular complementation group is known. In theory, the number of complementation groups is equal to the number of

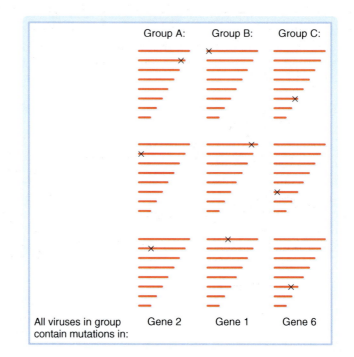

FIGURE 3.2 Complementation groups in influenza.
Complementation groups in influenza (or other viruses) all contain a mutation in the same virus gene, preventing the rescue of another mutant virus genome from the same complementation group.

genes in the virus genome. In practice, there are usually fewer complementation groups than genes, as mutations in some genes are always lethal and
other genes are nonessential and therefore cannot be scored in this type of
test. Complementation can be asymmetric—that is, only one of the mutant
viruses is able to replicate. This can be an absolute or a partial restriction.
When complementation occurs naturally, it is usually the case that a
replication-competent wild-type virus rescues a replication-defective mutant.
In these cases, the wild-type is referred to as a "helper virus," such as in the
case of defective transforming retroviruses containing oncogenes (see
Figure 3.3 and Chapter 7). Recombination is the physical interaction of virus
genomes during superinfection that results in gene combinations not present
in either parent. There are three mechanisms by which this can occur,
depending on the organization of the virus genome:

- **Intramolecular recombination via strand breakage and religation:**
  This process occurs in all DNA viruses and in RNA viruses that replicate
  via a DNA intermediate. It is believed to be caused by cellular enzymes,
  as no virus mutants with specific recombination defects have been
  isolated.
- **Intramolecular recombination by "copy-choice":** This process occurs
  in RNA viruses, probably by a mechanism in which the virus
  polymerase switches template strands during genome synthesis. There
  are cellular enzymes that could be involved (e.g., splicing enzymes),
  but this is unlikely and the process is thought to occur essentially as a

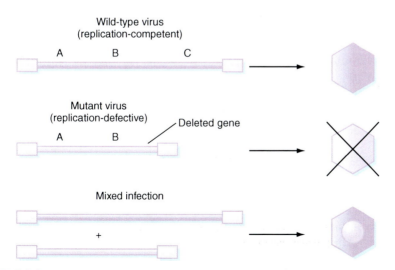

FIGURE 3.3 Helper viruses.
Helper viruses are replication-competent viruses that are capable of rescuing replication-defective
genomes in a mixed infection, permitting their multiplication and spread.

random event. Defective-interfering (DI) particles in RNA virus infections are frequently generated in this way (see Chapter 6).

- **Reassortment:** In viruses with segmented genomes, the genome segments can be randomly shuffled during superinfection. Progeny viruses receive (at least) one of each of the genome segments, but probably not from a single parent. For example, influenza virus has eight genome segments; therefore, in a mixed infection, there could be $2^8 = 256$ possible progeny viruses. Packaging mechanisms in these viruses are not well understood (see Chapter 2) but may be involved in generating reassortants.

In intramolecular recombination, the probability that breakage-reunion or strand-switching will occur between two markers (resulting in recombination) is proportional to the physical distance between them. Pairs of genetic markers can be arranged on a linear map with distances measured in "map units" (i.e., percentage recombination frequency). In reassortment, the frequency of recombination between two markers is either very high (indicating that the markers are on two different genome segments) or comparatively low (which means that they are on the same segment). This is because the frequency of reassortment usually swamps the lower background frequency that is due to intermolecular recombination between strands.

Reactivation is the generation of infectious (recombinant) progeny from non-infectious parental virus genomes. This process has been demonstrated *in vitro* and may be important *in vivo*. For example, it has been suggested that the rescue of defective, long-dormant HIV proviruses during the long clinical course of acquired immune deficiency syndrome (AIDS) may result in increased antigenic diversity and contribute to the pathogenesis of the disease. Recombination occurs frequently in nature; for example, influenza virus reassortment has resulted in worldwide epidemics (pandemics) that have killed millions of people (Chapter 6). This makes these genetic interactions of considerable practical interest and not merely an academic concern.

NONGENETIC INTERACTIONS BETWEEN VIRUSES

A number of nongenetic interactions between viruses occur which can affect the outcome and interpretation of the results of genetic crosses. Eukaryotic cells have a diploid genome with two copies of each chromosome, each bearing its own allele of the same gene. The two chromosomes may differ in allelic markers at many loci. Among viruses, only retroviruses are truly diploid, with two complete copies of the entire genome, but some DNA viruses, such as herpesviruses, have repeated sequences and are therefore partially heterozygous. In a few (mostly enveloped) viruses, aberrant packaging of multiple genomes may occasionally result in multiploid particles that are heterozygous (e.g., up to 10% of Newcastle disease virus particles). This process is known as **hetero-zygosis** and can contribute to the genetic complexity of virus populations.

Another nongenetic interaction between viruses that is commonly seen is interference. This process results from the resistance to **superinfection** by a virus observed in cells already infected by another virus. Homologous inter-ference (i.e., against the same virus) often results from the presence of DI particles which compete for essential cell components and block replication. However, interference can also result from other types of mutation (e.g., dominant temperature-sensitive mutations) or by sequestration of virus

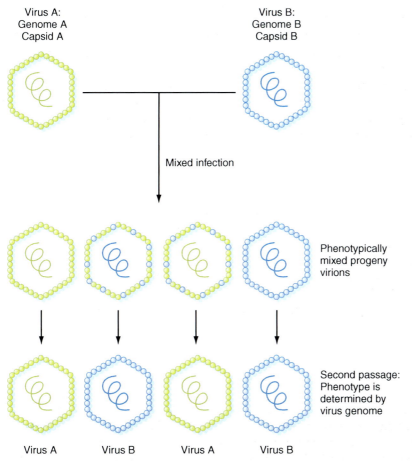

FIGURE 3.4 **Phenotypic mixing.**
Phenotypic mixing occurs in mixed infections, resulting in genetically unaltered virus particles that have some of the properties of the other parental type due to sharing a capsid.

receptors due to the production of **virus-attachment proteins** by viruses already present within the cell (e.g., in the case of avian retroviruses).

Phenotypic mixing can vary from extreme cases, where the genome of one virus is completely enclosed within the **capsid** or **envelope** of another (**pseudotyping**), to more subtle cases where the capsid/envelope of the progeny contains a mixture of proteins from both viruses. This mixing gives the progeny virus the phenotypic properties (e.g., cell **tropism**) dependent on the proteins incorporated into the particle, without any genetic change. Subsequent generations of viruses inherit and display the original parental phenotypes. This process can occur easily in viruses with naked capsids (nonenveloped) which are closely related (e.g., different strains of enteroviruses) or in enveloped viruses, which need not be related to one another (Figure 3.4). In this latter case, the phenomenon is due to the nonspecific incorporation of different virus glycoproteins into the envelope, resulting in a mixed phenotype. Rescue of replication-defective transforming retroviruses by a helper virus is a form of pseudotyping. Phenotypic mixing has proved to be a very useful tool to examine biological properties of viruses. Vesicular stomatitis virus (VSV) readily forms pseudotypes containing retrovirus envelope glycoproteins, giving a plaque-forming virus with the properties of VSV but with the cell tropism of the retrovirus. This trick has been used to study the cell tropism of HIV and other retroviruses.

SMALL DNA GENOMES

Bacteriophage M13 has already been mentioned in Chapter 2. The genome of this phage consists of 6.4 kb of single-stranded, positive-sense, circular DNA and encodes 10 genes. Unlike most **icosahedral** virions, the filamentous M13 **capsid** can be expanded by the addition of further protein subunits, so the genome length can be increased by the addition of extra sequences in the nonessential intergenic region without becoming incapable of being packaged into the capsid. In other bacteriophages, the genome packaging limits are more rigid. For example in phage λ, only DNA of between approximately 95 and 110% (approximately 46−54 kbp) of the normal genome size (49 kbp) can be packaged into the virus particle. Not all bacteriophages have such simple genomes as M13. The genome of phage T4 is about 160 kbp double-stranded DNA, and that of phage φKZ of *Pseudomonas aeruginosa* 280 kbp.

T4 and λ also illustrate another common feature of linear virus genomes—the importance of the sequences present at the ends of the genome. In the case of phage λ, the substrate packaged into the phage heads during assembly consists of long repetitive strings (concatemers) of phage DNA that are

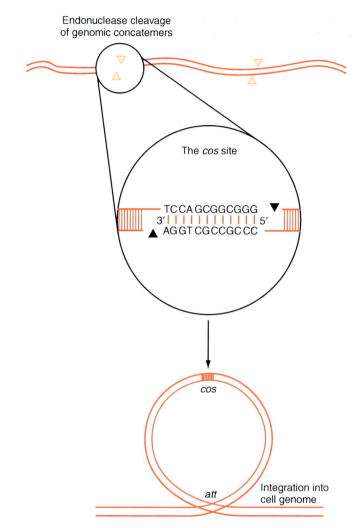

Endonuclease cleavage
of genomic concatemers

The *cos* site

TCCAGCGGCGGG
3' | | | | | | | | | | | | 5'
AGGTCGCCGCCC

cos

att   Integration into
                     cell genome

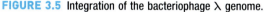

FIGURE 3.5 Integration of the bacteriophage λ genome.
The cohesive sticky ends of the *cos* site in the bacteriophage λ genome are ligated together in newly infected cells to form a circular molecule. Integration of this circular form into the *E. coli* chromosome occurs by specific recognition and cleavage of the *att* site in the phage genome.

produced during the later stages of genome replication. The DNA is reeled in by the phage head, and when a complete genome length has been incorporated, the DNA is cleaved at a specific sequence by a phage-coded endonuclease (Figure 3.5). This enzyme leaves a 12-bp 5' overhang on the end of each of the cleaved strands, known as the *cos* site. Hydrogen bond formation between these "sticky ends" can result in the formation of a circular molecule. In a newly infected cell, the gaps on either side of the *cos* site are closed

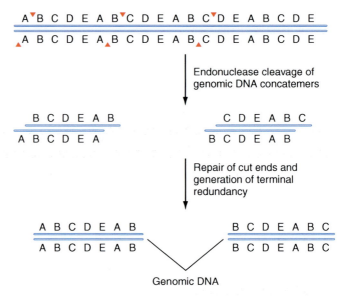

FIGURE 3.6 Terminal redundancy.
Terminal redundancy in the bacteriophage T4 genome results in the reiteration of some genetic information.

by DNA ligase, and it is this circular DNA that undergoes vegetative replication or integration into the bacterial chromosome. Phage T4 illustrates another molecular feature of certain linear virus genomes—**terminal redundancy**. Replication of the T4 genome also produces long concatemers of DNA. These are cleaved by a specific endonuclease, but unlike the λ genome the lengths of DNA incorporated into the particle are somewhat longer than a complete genome length (Figure 3.6). Some genes are repeated at each end of the genome, and the DNA packaged into the phage particles contains repeated information. These examples show that bacteriophage genomes are neither necessarily small nor simple.

As further examples of small DNA genomes, consider those of two groups of animal viruses: the parvoviruses and polyomaviruses. Parvovirus genomes are linear, nonsegmented, single-stranded DNA of about 5 kb. Parvovirus genomes are negative-sense, but some parvoviruses package equal amounts of (+) and (−) strands into **virions**. These are very small genomes, and even the replication-competent parvoviruses contain only two genes: *rep*, which encodes proteins involved in transcription, and *cap*, which encodes the coat proteins. However, the expression of these genes is rather complex, resembling the pattern seen in adenoviruses, with multiple **splicing** patterns seen for each gene (Chapter 5). The ends of the genome have palindromic sequences of about 115 nt, which form "hairpins" (Figure 3.7). These

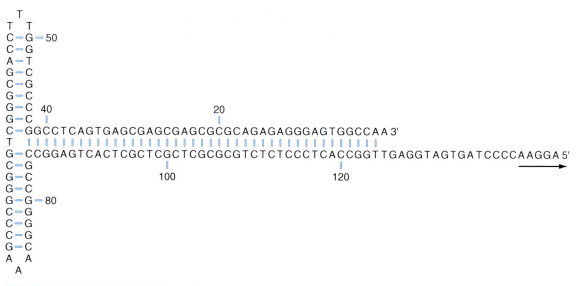

FIGURE 3.7 Parvovirus hairpin sequences.
Palindromic sequences at the ends of parvovirus genomes result in the formation of hairpin structures involved in the initiation of replication.

structures are essential for the initiation of genome replication, again emphasizing the importance of the sequences at the ends of the genome, and also determine which strand is packaged into virus particles.

The genomes of polyomaviruses are double-stranded, circular DNA molecules of approximately 5 kbp. The architecture of the polyomavirus genome (i.e., number and arrangement of genes and function of the regulatory signals and systems) has been studied in great detail. Within the particles, the virus DNA assumes a supercoiled form and is associated with four cellular histones: H2A, H2B, H3, and H4 (see Chapter 2). The genomic organization of these viruses has evolved to pack the maximum information (six genes) into minimal space (5 kbp). This has been achieved by the use of both strands of the genome DNA and overlapping genes (Figure 3.8). VP1 is encoded by a dedicated ORF, but the VP2 and VP3 genes overlap so that VP3 is contained within VP2. The origin of replication is surrounded by noncoding regions which control transcription. Polyomaviruses also encode "T-antigens," which are proteins that can be detected by sera from animals bearing polyomavirus-induced tumors. These proteins bind to the origin of replication and show complex activities in that they are involved both in DNA replication and in the transcription of virus genes. This topic is discussed further in Chapter 7.

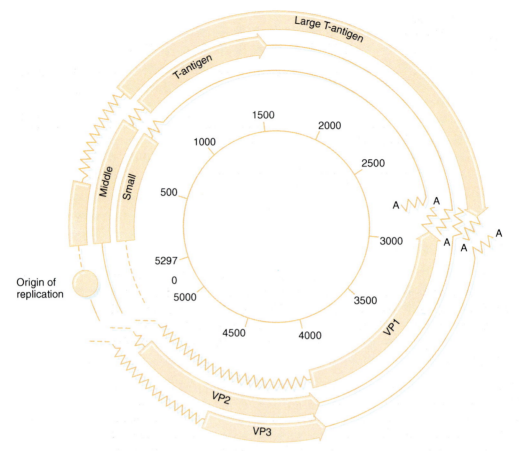

FIGURE 3.8 Polyomavirus genome.
The complex organization of the polyomavirus genome results in the compression of much genetic information into a relatively short sequence.

LARGE DNA GENOMES

A number of virus groups have double-stranded DNA genomes of consider-able size and complexity. In many respects, these viruses are genetically simi-lar to the host cells that they infect. The genomes of adenoviruses consist of linear, double-stranded DNA of 30−38 kbp, the precise size of which varies between different adenoviruses. These genomes contain 30−40 genes (Figure 3.9). The terminal sequence of each DNA strand is an inverted repeat of 100−140 bp, and the denatured single strands can form "panhandle" structures. These structures are important in DNA replication, as is a 55-kDa protein known as the terminal protein that is covalently attached to the 5′ end of each strand. During genome replication, this protein acts as a primer,

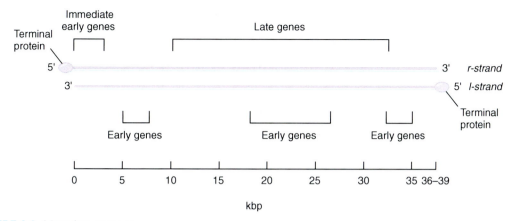

FIGURE 3.9 Adenovirus genomes.
Organization of the adenovirus genome.

initiating the synthesis of new DNA strands. Although adenovirus genomes are considerably smaller than those of herpesviruses, the expression of the genetic information is rather more complex. Clusters of genes are expressed from a limited number of shared promoters. Multiply spliced mRNAs and alternative splicing patterns are used to express a variety of polypeptides from each promoter (see Chapter 5).

The *Herpesviridae* is a large family containing more than 100 different members, at least one for most animal species that have been examined to date. There are eight human herpesviruses, all of which share a common overall genome structure but which differ in the fine details of genome organization and at the level of nucleotide sequence. The family is divided into three subfamilies, based on their nucleotide sequence and biological properties (Table 3.1). Herpesviruses have large genomes composed of up to 235 kbp of linear, double-stranded DNA and correspondingly large and complex virus particles containing about 35 virion polypeptides. All encode a variety of enzymes involved in nucleic acid metabolism, DNA synthesis, and protein processing (e.g., protein kinases). The different members of the family are widely separated in terms of genomic sequence and proteins, but all are similar in terms of structure and genome organization (Figure 3.10A). Some but not all herpesvirus genomes consist of two covalently joined sections, a unique long (U_L) and a unique short (U_S) region, each bounded by inverted repeats. The repeats allow structural rearrangements of the unique region. This arrangement allows these genomes to exist as a mixture of four structural isomers, all of which are functionally equivalent (Figure 3.10B). Herpesvirus genomes also contain multiple repeated sequences and, depending on the number of these, the genome size of different isolates of a particular virus can vary by up to 10 kbp. The prototype member of the family is herpes simplex virus (HSV), whose genome consists of approximately

Table 3.1 Human Herpesviruses	
Alphaherpesvirinae	
Latent infections in sensory ganglia; genome size 120—180 kbp	
Simplexvirus	Human herpesviruses 1 and 2 (HSV-1, HSV-2)
Varicellovirus	Human herpesvirus 3 (VZV)
Betaherpesvirinae	
Restricted host range; genome size 140—235 kbp	
Cytomegalovirus	Human herpesvirus 5 (HCMV)
Roseolovirus	Human herpesviruses 6 and 7 (HHV-6, HHV-7)
Gammaherpesvirinae	
Infection of lymphoblastoid cells; genome size 105—175 kbp	
Lymphocryptovirus	Human herpesvirus 4 (EBV)
Rhadinovirus	Human herpesvirus 8 (HHV-8)

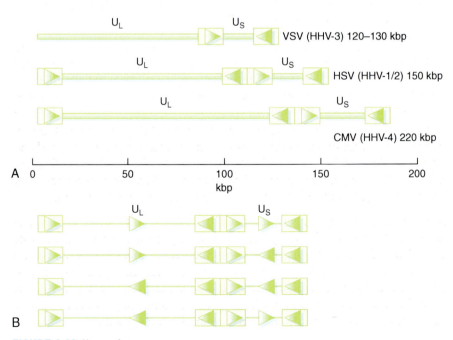

FIGURE 3.10 Herpesvirus genomes.
(A) Some herpesvirus genomes consist of two covalently joined sections, U_L and U_S, each bounded by inverted repeats. (B) This organization permits the formation of four different isomeric forms of the genome.

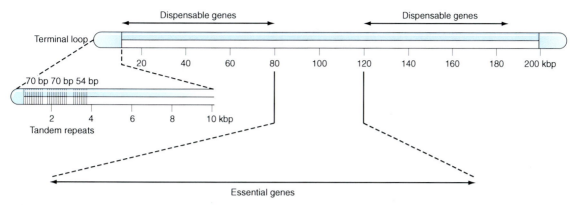

FIGURE 3.11 Poxvirus genome organization.
In these large and complex genomes, essential genes are located in the central region of the genome. Genes that are dispensable for replication in culture are located closer to the ends of the genome while sequences at the end of the strand contain many sequence repeats important for genome replication.

152 kbp of double-stranded DNA, the complete nucleotide sequence of which has now been determined. This virus contains about 80 genes, densely packed and with overlapping reading frames. Each gene is expressed from its own pro-moter (see adenovirus discussion above).

Poxvirus genomes are linear structures ranging in size from 140 to 290 kbp. As with the herpesviruses, each gene tends to be expressed from its own promoter. Characteristically, the central regions of poxvirus genomes tend to be highly conserved and to contain essential genes which are essential for replication in culture, while the outer regions of the genome are more variable in sequence and at least some of the genes located here are dispensable (Figure 3.11). In contrast, the non-coding nucleic acid structures at the ends of the genome are highly conserved and vital for replication. There are no free ends to the linear genome because these are closed by "hairpin" arrangements. Adjacent to the ends of the genome are other noncoding sequences which play vital roles in replication (see Chapter 4).

In the last few years, viruses with even larger DNA genomes have been discovered. At the time of writing, the largest virus genomes known are those of the recently discovered Pandoraviruses, double-stranded DNA genomes of up to 2.8 Mbp—approaching the size of the simplest eukaryotic genomes—encapsidated in micron-sized amphora-shaped particles. But the upper size limit for virus genomes is increasing all the time and likely to keep doing so for some years as new methods of virus discovery are applied.

POSITIVE-STRAND RNA VIRUSES

The ultimate size of single-stranded RNA genomes is limited by the relatively fragile nature of RNA and the tendency of long strands to break. In addition,

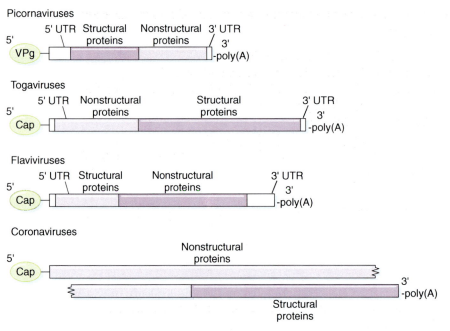

FIGURE 3.12 Genomic organization of positive-stranded RNA viruses.
This diagram illustrates some of the differences and similar features between different families of positive-stranded RNA viruses.
Differences in patterns of gene expression are discussed in Chapter 5.

RNA genomes tend to have higher mutation rates than those composed of DNA because they are copied less accurately, although the virus-encoded RNA-dependent RNA polymerases responsible for the replication of these genomes do have some repair mechanisms. These reasons tend to drive RNA viruses toward smaller genome sizes. Single-stranded RNA genomes vary in size from those of coronaviruses, which are approximately 30 kb long, to those of bacteriophages such as MS2 and Qβ, at about 3.5 kb, although members of distinct families, most positive-sense RNA viruses of vertebrates share common features in terms of the biology of their genomes. In particular, purified positive-sense virus RNA is directly infectious when applied to susceptible host cells in the absence of any virus proteins (although it is about one million times less infectious than virus particles). On examining the features of these virus families, although the details of genomic organization vary, some repeated themes emerge (Figure 3.12).

Picornaviruses

The picornavirus genome consists of one single-stranded, positive-sense RNA molecule of between 7.2 kb in human rhinoviruses and 8.5 kb in foot-and-mouth disease viruses, containing a number of features conserved in all picornaviruses:

- There is a long (600−1200 nt) untranslated region (UTR) at the 5′ end which is important in translation, virulence, and possibly encapsidation, as well as a shorter 3′ UTR (50−100 nt) that is necessary for (−)strand synthesis during replication.
- The 5′ UTR contains a "clover-leaf" secondary structure known as the internal ribosomal entry site (IRES) (Chapter 5).
- The rest of the genome encodes a single polyprotein of between 2,100 and 2,400 amino acids.

Both ends of the genome are modified—the 5′ end by a covalently attached small, basic protein VPg (23 amino acids) and the 3′ end by polyadenylation.

Togaviruses

The togavirus genome is comprised of single-stranded, positive-sense, non-segmented RNA of approximately 11.7 kb. It has the following features:

- It resembles cellular mRNAs in that it has a 5′ methylated cap and 3′ poly(A) sequences.
- Gene expression is achieved by two rounds of translation, producing first nonstructural proteins encoded in the 5′ part of the genome and later structural proteins from the 3′ part.

Flaviviruses

The flavivirus genome is a single-stranded, positive-sense RNA molecule of about 10.5 kb with the following features:

- It has a 5′ methylated cap, but in most cases the RNA is not polyadenylated at the 3′ end.
- Genetic organization differs from that of the togaviruses in that the structural proteins are encoded in the 5′ part of the genome and nonstructural proteins in the 3′ part.

Expression is similar to that of the picornaviruses, involving the production of a polyprotein.

Coronaviruses

The coronavirus genome consists of nonsegmented, single-stranded, positive-sense RNA, approximately 27−30 kb long, which is the longest of any RNA virus. It also has the following features:

- It has a 5′ methylated cap and 3′ poly(A), and the vRNA functions directly as mRNA.

- The 5′ 20-kb segment of the genome is translated first to produce a virus polymerase, which then produces a full-length negative-sense strand. This is used as a template to produce mRNA as a nested set of transcripts, all with an identical 5′ nontranslated leader sequence of 72 nt and coincident 3′ polyadenylated ends.
- Each mRNA is monocistronic, the genes at the 5′ end being translated from the longest mRNA and so on. These unusual cytoplasmic structures are produced not by splicing (posttranscriptional modification) but by the polymerase during transcription.

Positive-Sense RNA Plant Viruses

The majority (but not all) of plant virus families have positive-sense RNA genomes. The genome of the tobamovirus tobacco mosaic virus (TMV) is a well-studied example (Figure 3.13):

- The TMV genome is a 6.4-kb RNA molecule that encodes four genes.
- There is a 5′ methylated cap, and 3′ end of the genome contains extensive secondary structure but no poly(A) sequences.

Expression is reminiscent of but distinct from that of togaviruses, producing nonstructural proteins by direct translation of the ORF encoded in the 5′ part of the genome and the virus coat protein and further nonstructural proteins from two subgenomic RNAs encoded by the 3′ part. The similarities and differences between genomes in this class will be considered further in the discussion of virus evolution below and in Chapter 5.

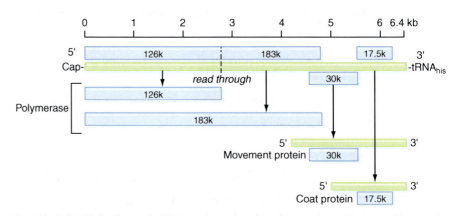

FIGURE 3.13 Organization of the TMV genome.
This positive-sense RNA plant virus expresses several genes via subgenomic messenger RNAs (see Chapter 5).

NEGATIVE-STRAND RNA VIRUSES

Viruses with **negative-sense** RNA genomes are more diverse than the positive-stranded RNA viruses discussed above. Because of the difficulties of gene expression and genome replication, they tend to have larger genomes encoding more genetic information. Because of this, segmentation is a common, although not universal, feature of such viruses (Figure 3.14). None of these genomes is infectious as purified RNA. Although a gene encoding an RNA-dependent RNA polymerase has been found in some eukaryotic cells, most uninfected cells do not contain enough RNA-dependent RNA polymerase activity to support virus replication, and, because the negative-sense genome cannot be translated as mRNA without the virus polymerase packaged in each particle, these genomes are effectively inert. A few of the viruses described in this section are not strictly negative-sense but are **ambisense**, as

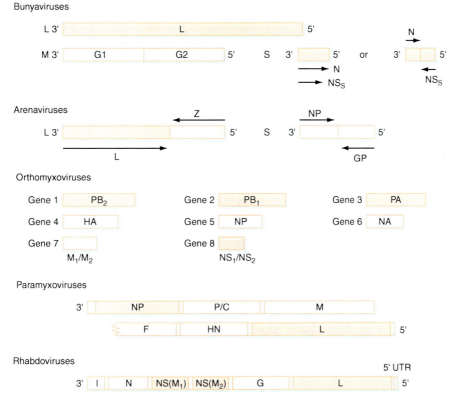

FIGURE 3.14 Genome organization of negative-stranded RNA viruses.
The fundamental distinction in the negative-strand RNA viruses is between those viruses with segmented genomes and those with nonsegmented genomes.

they are part negative-sense and part positive-sense. Ambisense coding strategies occur in both plant viruses (e.g., the *Tospovirus* genus of the bunyaviruses, and tenuiviruses such as rice stripe virus) and animal viruses (the *Phlebovirus* genus of the bunyaviruses, and arenaviruses).

BOX 3.2 CAN'T MAKE YOUR MIND UP? DO BOTH!

Ambisense virus genomes contain at least one RNA segment which is part positive and part of negative-sense—in the same molecule. In spite of this, genetically they have more in common with negative-strand viruses than positive RNA viruses. But why on Earth would any virus bother with such a complicated gene expression strategy? In general, it is more difficult for RNA viruses to control gene expression that it is for DNA viruses to upregulate and downregulate individual gene products. Most ambisense viruses can replicate in a range of hosts, such as mammals and insects or insects and plants. In their vector or reservoir host, infection is usually asymptomatic. However, in another host, multiplication of the virus can be lethal. Having two different strategies for gene expression may help them to successfully span this divide.

Bunyaviruses

Members of the *Bunyaviridae* have single-stranded, negative-sense, segmented RNA genomes with the following features:

- The genome is comprised of three molecules: L (8.5 kb), M (5.7 kb), and S (0.9 kb).
- All three RNA species are linear, but in the **virion** they appear circular because the ends are held together by base-pairing. The three segments are not present in virus preparations in equimolar amounts.
- In common with all negative-sense RNAs, the 5′ ends are not capped and the 3′ ends are not polyadenylated.

Members of the *Phlebovirus* and *Tospovirus* genera differ from the other three genera in the family (*Bunyavirus*, *Nairovirus*, and *Hantavirus*) in that genome segment S is rather larger and the overall genome organization is different—**ambisense** (i.e., the 5′ end of each segment is positive-sense, but the 3′ end is negative-sense). The *Tospovirus* genus also has an ambisense coding strategy in the M segment of the genome.

Arenaviruses

Arenavirus genomes consist of linear, single-stranded RNA. There are two genome segments: L (5.7 kb) and S (2.8 kb). Both segments have an ambisense organization, as in the Bunyaviruses.

Orthomyxoviruses

See discussion of segmented genomes (in the next section).

Paramyxoviruses

Members of the *Paramyxoviridae* have nonsegmented negative-sense RNA of 15−16 kb. Typically, six genes are organized in a linear arrangement (3′−NP−P/C/V−M−F−HN−L−5′) separated by repeated sequences: a polyadenylation signal at the end of the gene, an intergenic sequence (GAA), and a translation start signal at the beginning of the next gene.

Rhabdoviruses

Viruses of the *Rhabdoviridae* have nonsegmented, negative-sense RNA of approximately 11 kb. There is a leader region of approximately 50 nt at the 3′ end of the genome and a 60 nt UTR at the 5′ end of the vRNA. Overall, the genetic arrangement is similar to that of paramyxoviruses, with a conserved polyadenylation signal at the end of each gene and short intergenic regions between the five genes.

SEGMENTED AND MULTIPARTITE VIRUS GENOMES

There is sometimes confusion between these two types of genome structure. Segmented virus genomes are those that are divided into two or more physically separate molecules of nucleic acid, all of which are then packaged into a single virus particle. In contrast, although multipartite genomes are also segmented, each genome segment is packaged into a separate virus particle. These discrete particles are structurally similar and may contain the same component proteins, but they often differ in size depending on the length of the genome segment packaged. In one sense, multipartite genomes are, of course, segmented, but this is not the strict meaning of these terms as they will be used here.

Segmentation of the virus genome has a number of advantages and disadvantages. There is an upper limit to the size of a nonsegmented virus genome which results from the physical properties of nucleic acids, particularly the tendency of long molecules to break due to shear forces (and, for each particular virus, the length of nucleic acid that can be packaged into the capsid). The problem of strand breakage is particularly relevant for single-stranded RNA, which is less chemically stable than double-stranded DNA. The longest single-stranded RNA genomes are those of the coronaviruses at approximately 30 kb, but the longest double-stranded DNA virus genomes are considerably longer (e.g., Pandoravirus at up to 2.8 Mbp). Physical breakage of the genome results in biological inactivation, as it cannot be completely transcribed, translated, or replicated. Segmentation means that the virus avoids "having all its eggs in one basket" and also reduces the probability of breakages due to shearing, thus increasing the total potential coding capacity of the entire genome. The disadvantage of segmentation is that all the individual genome segments must be packaged into each virus particle or the virus will be defective as a result of loss of genetic information. In general, it is not understood how this control of packaging is achieved.

Separating the genome segments into different particles (the multipartite strategy) removes the requirement for accurate sorting but introduces a new problem in that all the discrete virus particles must be taken up by a single host cell to establish a productive infection. This is perhaps the reason why multipartite viruses are only found in plants. Many of the sources of infection by plant viruses, such as inoculation by sapsucking insects or after physical damage to tissues, result in a large inoculum of infectious virus particles, providing opportunities for infection of an initial cell by more than one particle.

The genetics of segmented genomes are essentially the same as those of nonsegmented genomes, with the addition of the reassortment of segments. Reassortment can occur whether the segments are packaged into a single particle or are in a multipartite configuration. Reassortment is a powerful means of achieving rapid generation of genetic diversity; this could be another possible reason for its evolution. Segmentation of the genome also has implications for the partition of genetic information and the way in which it is expressed, which will be considered further in Chapter 5.

Table 3.2 Influenza Virus Genome Segments

Segment	Size (nt)	Polypeptides	Function (Location)
1	2,341	PB_2	Transcriptase: cap binding
2	2,341	PB_1	Transcriptase: elongation
3	2,233	PA	Transcriptase: (?)
4	1,778	HA	Hemagglutinin
5	1,565	NP	Nucleoprotein: RNA binding; part of transcriptase complex
6	1,413	NA	Neuraminidase
7	1,027	M_1	Matrix protein: major component of virion
		M_2	Integral membrane protein—ion channel
8	890	NS_1	Nonstructural (nucleus): function unknown
		NS_2	Nonstructural (nucleus + cytoplasm): function unknown

To understand the complexity of these genomes, consider the organization of a segmented virus genome (influenza A virus) and a multipartite genome (geminivirus). The influenza virus genome is composed of eight segments (in influenza A and B strains; seven in influenza C) of single-stranded, negative-sense RNA (Table 3.2). The identity of the proteins encoded by each genome segment were determined originally by genetic analysis of the electrophoretic mobility of the individual segments from reassortant viruses and by analysis of a large number of mutants covering all eight segments. The eight segments have common nucleotide sequences at the 5′ and 3′ ends (Figure 3.15) which are necessary for replication of the genome (Chapter 4). These sequences are complementary to one another, and, inside the particle, the ends of the genome segments are held together by base-pairing and form a panhandle structure that again is believed to be involved in replication. The RNA genome segments are not packaged as naked nucleic acid but in association with the gene 5 product, the nucleoprotein, and are visible in electron micrographs as helical structures. Here there is a paradox. Biochemically and genetically, each genome segment behaves as an individual, discrete entity; however, in electron micrographs of influenza virus particles disrupted with nonionic detergents, the nucleocapsid has the physical appearance of a single, long helix. Clearly, there is some interaction between the genome segments and it is this which explains the ability of influenza virus particles to select and package the genome segments within each particle with a surprisingly low error rate, considering the difficulty of the task (Chapter 2).

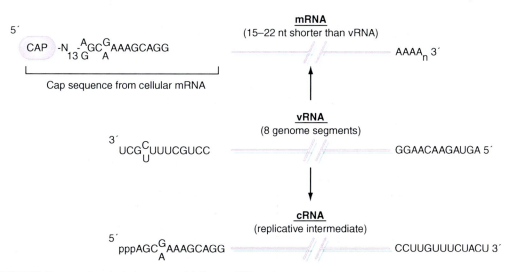

FIGURE 3.15 Common terminal sequences of influenza RNAs.
Influenza virus genome segments are a classic example of how sequences at the ends of liner virus genomes are crucial for gene expression and for replication.

In many tropical and subtropical parts of the world, geminiviruses are important plant pathogens. Geminiviruses are divided into three genera based on their host plants (monocotyledons or dicotyledons) and insect vectors (leafhoppers or whiteflies). In the *Mastrevirus* and *Curtovirus* genera, the genome consists of a single-stranded DNA molecule of approximately 2.7 kb. The DNA packaged into these **virions** has been arbitrarily designated as positive-sense, although both the positive-sense and negative-sense strands found in infected cells contain protein-coding sequences. The genome of geminiviruses in the genus *Begmovirus* is bipartite and consists of two circular, single-stranded DNA molecules, each of which is packaged into a separate particle (Figure 3.16). Both of the strands comprising the genome are approximately 2.7 kb long and differ from one another completely in nucleotide sequence, except for a shared 200-nt noncoding sequence involved in DNA replication. The two genomic DNAs are packaged into entirely separate **capsids**. Because a **productive infection** requires both parts of the genome, it is necessary for a minimum of two virus particles bearing one copy of each of the genome segments to infect a new host cell. Although geminiviruses do not multiply in the tissues of their insect vectors (**nonpropagative transmission**), a sufficiently large amount of virus is ingested and subsequently deposited onto a new host plant to favor such **superinfections**. An even more

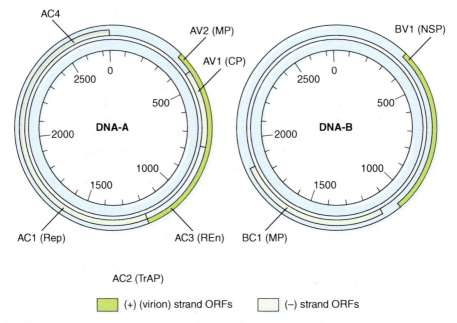

FIGURE 3.16 Bipartite geminivirus genome.
Organization and protein-coding potential of a bipartite begmovirus (*Geminiviridae*) genome. Proteins encoded by the positive-sense virion strand are named "A" or "B" depending on which of the two genome components they are located on, or "V" or "C" depending on whether they are encoded by the positive-sense virion strand of the negative-sense complementary strand.

extreme example is that of the Nanoviruses, multipartite viruses with up to 11 genome segments.

Both of these examples show a high density of coding information. In influenza virus, genes 7 and 8 both encode two proteins in overlapping reading frames. In geminiviruses, both strands of the virus DNA found in infected cells contain coding information, some of which is present in overlapping reading frames. It is possible that this high density of genetic information is the reason why these viruses have resorted to divided genomes, in order to regulate the expression of this information (see Chapter 5).

REVERSE TRANSCRIPTION AND TRANSPOSITION

The first successes of molecular biology were the discovery of the double-helix structure of DNA and in understanding of the language of the genetic code. The importance of these findings does not lie in their chemistry but in their importance in allowing predictions to be made about the fundamental nature of living organisms. The confidence that flowed from these early triumphs resulted in the development of a grand universal theory, called the "central dogma of molecular biology"—namely, that all cells (and hence viruses) work on a simple organizing principle: the unidirectional flow of information from DNA, through RNA, into proteins. In the mid-1960s, however, there were rumblings that life might not be so simple.

In 1963, Howard Temin showed that the replication of retroviruses, whose particles contain RNA genomes, was inhibited by actinomycin D, an antibiotic that binds only to DNA. The replication of other RNA viruses is not inhibited by this drug. So pleased was the scientific community with an all-embracing dogma that these facts were largely ignored until 1970, when Temin and David Baltimore simultaneously published the observation that retrovirus particles contain an RNA-dependent DNA polymerase: reverse transcriptase. This finding alone was important enough, but like the earlier conclusions of molecular biology, it has subsequently had reverberations for the genomes of all organisms, and not merely a few virus families. It is now known that retrotransposons with striking similarities to retrovirus genomes form a substantial part of the genomes of all higher organisms, including humans. Earlier ideas of genomes as constant, stable structures have been replaced with the realization that they are, in fact, dynamic and rather fluid entities.

The concept of transposable genetic elements—specific sequences that are able to move from one position in the genome to another—was put forward by Barbara McClintock in the 1940s. Such transposons fall into two groups:

- Simple transposons, which do not undergo reverse transcription and are found in prokaryotes (e.g., the genome of enterobacteria phage Mu).

■ Retrotransposons, which closely resemble retrovirus genomes and are bounded by long direct repeats (long terminal repeats, or LTRs); these move by means of a transcription/reverse transcription/integration mechanism and are found in eukaryotes (the *Metaviridae* and *Pseudoviridae*).

Both types show a number of similar properties:

■ They are believed to be responsible for a high proportion of apparently spontaneous mutations.
■ They promote a wide range of genetic rearrangements in host cell genomes, such as deletions, inversions, duplications, and translocations of the neighboring cellular DNA.
■ The mechanism of insertion generates a short (3−13 bp) duplication of the DNA sequence on either side of the inserted element.
■ The ends of the transposable element consist of inverted repeats, 2−50 bp long.
■ Transposition is often accompanied by replication of the element— necessarily so in the case of retrotransposons, but this also often occurs with prokaryotic transposition.

Transposons control their own transposition functions, encoding proteins that act on the element in *cis* (affecting the activity of contiguous sequences on the same nucleic acid molecule) or in *trans* (encoding diffusible products acting on regulatory sites in any stretch of nucleic acid present in the cell). Bacteriophage Mu infects *E. coli* and consists of a complex, tailed particle containing a linear, double-stranded DNA genome of about 37 kb, with host-cell-derived sequences of between 0.5 and 2 kbp attached to the right-hand end of the genome (Figure 3.17). Mu is a temperate bacteriophage whose replication can proceed through two pathways; one involves integration of the genome into that of the host cell and results in lysogeny, and the other is lytic replication, which results in the death of the cell (see Chapter 5). Integration of the phage genome into that of the host bacterium occurs at random sites in the cell genome. Integrated phage genomes are

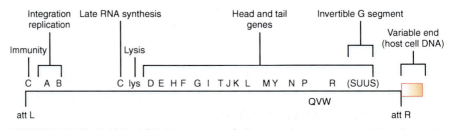

FIGURE 3.17 Bacteriophage Mu genome.
Organization of the phage Mu genome.

known as prophage, and integration is essential for the establishment of lysogeny. At intervals in bacterial cells lysogenic for Mu, the prophage undergoes transposition to a different site in the host genome. The mechanism leading to transposition is different from that responsible for the initial integration of the phage genome (which is conservative in that it does not involve replication) and is a complex process requiring numerous phage-encoded and host-cell proteins. Transposition is tightly linked to replication of the phage genome and results in the formation of a "co-integrate"—that is, a duplicate copy of the phage genome flanking a target sequence in which insertion has occurred. The original Mu genome remains in the same location where it first integrated and is joined by a second integrated genome at another site. (Not all prokaryotic transposons use this process; some, such as TN10, are not replicated during transposition but are excised from the original integration site and integrate elsewhere.) There are two consequences of such a transposition. First, the phage genome is replicated during this process (advantageous for the virus), and second, the sequences flanked by the two phage genomes (which form repeated sequences) are at risk of secondary rearrangements, including deletions, inversions, duplications, and translocations (possibly but not necessarily deleterious for the host cell).

The yeast Ty viruses are representative of a class of sequences found in yeast and other eukaryotes known as retrotransposons. Unlike enterobacteria phage Mu, such elements are not true viruses but do bear striking similarities to retroviruses. The genomes of most strains of *Saccharomyces cerevisiae* contain 30−35 copies of the Ty elements, which are around 6 kbp long and contain direct repeats of 245−371 bp at each end (Figure 3.18). Within this repeat sequence is a promoter which leads to the transcription of a terminally

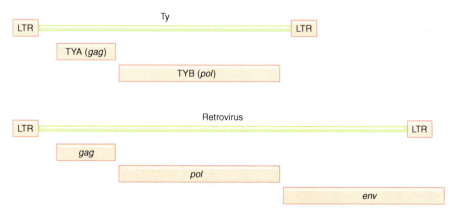

FIGURE 3.18 Retrotransposons and retroviruses.
The genetic organization of retrotransposons such as Ty (above) and retrovirus genomes (below) shows a number of similarities, including the presence of direct LTRs at either end.

redundant 5.6-kb mRNA. This contains two genes: TyA, which has homology to the *gag* gene of retroviruses, and TyB, which is homologous to the *pol* gene. The protein encoded by TyA is capable of forming a roughly spherical, 60-nm diameter, "virus-like particle" (VLP). The 5.6-kb RNA transcript can be incorporated into such particles, resulting in the formation of intracellular structures known as Ty-VLPs. Unlike true viruses these particles are not infectious for yeast cells, but if accidentally taken up by a cell they can carry out reverse transcription of their RNA content to form a double-stranded DNA Ty element, which can then integrate into the host cell genome.

The most significant difference between retrotransposons such as Ty, copia (a similar element found in *Drosophila melanogaster*), and retroviruses proper is the presence of an additional gene in retroviruses, *env*, which encodes an envelope glycoprotein (see Chapter 2). The envelope protein is responsible for receptor binding and has allowed retroviruses to escape the intracellular lifestyle of retrotransposons to form a true virus particle and propagate themselves widely by infection of other cells (Figure 3.18). Retrovirus genomes have four unique features:

- They are the only viruses that are truly diploid.
- They are the only RNA viruses whose genome is produced by cellular transcriptional machinery (without any participation by a virus-encoded polymerase).
- They are the only viruses whose genome requires a specific cellular RNA (tRNA) for replication.
- They are the only positive-sense RNA viruses whose genome does not serve directly as mRNA immediately after infection.

During the process of reverse transcription (Figure 3.19), the two single-stranded positive-sense RNA molecules that comprise the virus genome are converted into a double-stranded DNA molecule somewhat longer than the RNA templates due to the duplication of direct repeat sequences at each end—the LTRs (Figure 3.20). Some of the steps in reverse transcription have remained mysteries—for example, the apparent jumps that the polymerase makes from one end of the template strand to the other. In fact, these steps can be explained by the observation that complete conversion of retrovirus RNA into double-stranded DNA only occurs in a partially uncoated core particle and cannot be duplicated accurately *in vitro* with the reagents free in solution. This indicates that the conformation of the two RNAs inside the retrovirus nucleocapsid dictates the course of reverse transcription—the "jumps" are nothing of the sort, as the ends of the strands are probably held adjacent to one another inside the core.

Reverse transcription has important consequences for retrovirus genetics. First, it is a highly error-prone process, because reverse transcriptase does not carry

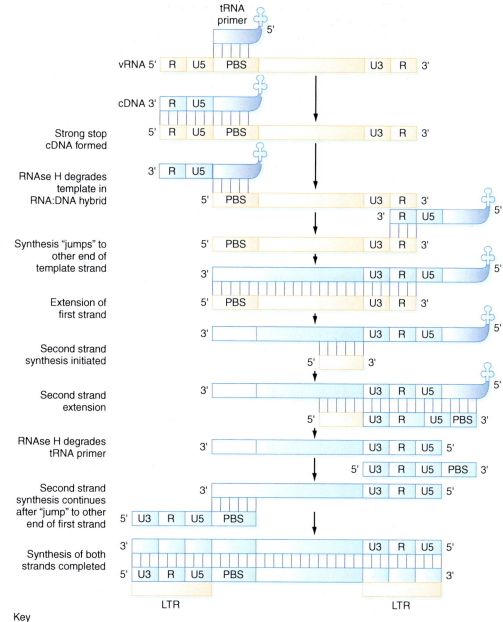

Key

☐ Virus RNA

☐ Newly synthesized cDNA

FIGURE 3.19 Reverse transcription.

Mechanism of reverse transcription of retrovirus RNA genomes, in which two molecules of RNA are converted into a single (terminally redundant) double-stranded DNA provirus.

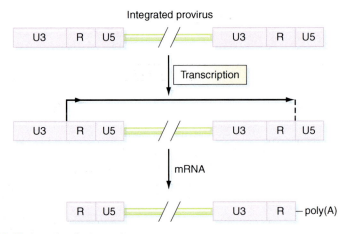

FIGURE 3.20 Long terminal repeats.
Generation of repeated information in retrovirus LTRs. In addition to their role in reverse transcription, these sequences contain important control elements involved in the expression of the virus genome, including a transcriptional promoter in the U3 region and polyadenylation signal in the R region.

out the proofreading functions performed by cellular DNA-dependent DNA polymerases. This results in the introduction of many mutations into retrovirus genomes and, consequently, rapid genetic variation. In addition, the process of reverse transcription promotes genetic **recombination**. Because two RNAs are packaged into each **virion** and used as the template for reverse transcription, recombination can and does occur between the two strands. Although the mechanism responsible for this is not clear, if one of the RNA strands differs from the other (e.g., by the presence of a mutation) and recombination occurs, then the resulting virus will be genetically distinct from either of the parental viruses.

After reverse transcription is complete, the double-stranded DNA migrates into the nucleus, still in association with virus proteins. The mature products of the *pol* gene are, in fact, a complex of polypeptides that include three distinct enzymatic activities: reverse transcriptase and RNAse H, which are involved in reverse transcription, and integrase, which catalyzes integration of virus DNA into the host cell **chromatin**, after which it is known as the **provirus** (Figure 3.21). Three forms of double-stranded DNA are found in retrovirus-infected cells following reverse transcription: linear DNA and two circular forms that contain either one or two LTRs. From the structure at the ends of the provirus, it was previously believed that the two-LTR circle was the form used for integration. In recent years, systems that have been developed to study the integration of retrovirus DNA *in vitro* show that it is the linear form that integrates. This discrepancy can be resolved by a model in

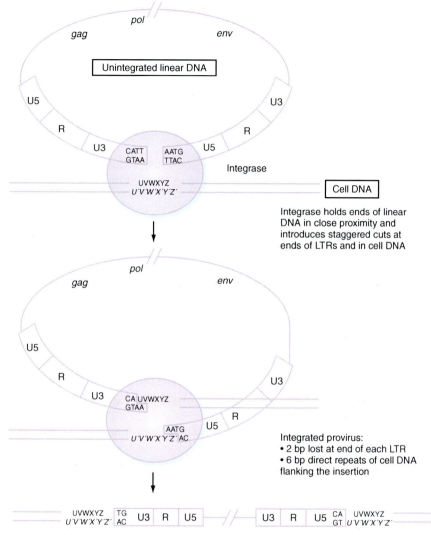

FIGURE 3.21 Retrovirus integration.
Mechanism of integration of retrovirus genomes into the host cell chromatin.

which the ends of the two LTRs are held in close proximity by the reverse transcriptase–integrase complex. The net result of integration is that 1–2 bp are lost from the end of each LTR and 4–6 bp of cellular DNA are duplicated on either side of the provirus. It is unclear whether there is any specificity regarding the site of integration into the cell genome. What is obvious is that there is no simple target sequence, but it is possible that there may be (numerous) regions or sites in the eukaryotic genome that are more likely to be integration sites than others.

Following integration, the DNA provirus genome becomes essentially a collection of cellular genes and is at the mercy of the cell for expression. There is no mechanism for the precise excision of integrated proviruses, some of which are known to have been fossilized in primate genomes through millions of years of evolution, although proviruses may sometimes be lost or altered by modifications of the cell genome. The only way out for the virus is transcription, forming what is essentially a full-length mRNA (minus the terminally redundant sequences from the LTRs). This RNA is the vRNA, and two copies are packaged into virions (Figure 3.20).

There are, however, two different groups of viruses whose replication involves reverse transcription. It is at this point that the difference between them becomes obvious. One strategy, used by retroviruses, culminates in the packaging of RNA into virions as the virus genome. The other, used by hepadnaviruses and caulimoviruses, switches the RNA and DNA phases of replication and results in DNA virus genomes inside virus particles. This is achieved by utilizing reverse transcription, not as an early event in replication as retroviruses do, but as a late step during formation of the virus particle.

Hepatitis B virus (HBV) is the prototype member of the family *Hepadnaviridae*. HBV virions are spherical, lipid-containing particles, 42−47 nm diameter, which contain a partially double-stranded ("gapped") DNA genome, plus an RNA-dependent DNA polymerase (i.e., reverse transcriptase) (Figure 3.22). Hepadnaviruses have very small genomes consisting of a negative-sense strand of 3.0−3.3 kb (varies between different hepadnaviruses) and a positive-sense strand of 1.7−2.8 kb (varies between different particles). On infection of cells, three major genome transcripts are produced: 3.5-, 2.4-, and 2.1-kb mRNAs. All have the same polarity (i.e., are transcribed from the same strand of the virus genome) and the same 3′ ends but have different 5′ ends (i.e., initiation sites). These transcripts are heterogeneous in size, and it is not completely clear which proteins each transcript encodes, but there are four known genes in the virus:

- C encodes the core protein.
- P encodes the polymerase.
- S encodes the three polypeptides of the surface antigen: pre-S1, pre-S2, and S (which are derived from alternative start sites).
- X encodes a transactivator of virus transcription (and possibly cellular genes?).

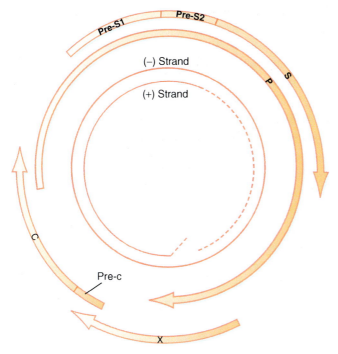

FIGURE 3.22 **HBV genome.**
Structure, organization, and proteins encoded by the HBV genome.

Closed circular DNA is found soon after infection in the nucleus of the cell and is probably the source of the above transcripts. This DNA is produced by repair of the gapped virus genome as follows:

1. Completion of the positive-sense strand.
2. Removal of a protein primer from the negative-sense strand and an oligoribonucleotide primer from the positive-sense strand.
3. Elimination of terminal redundancy at the ends of the negative-sense strand.
4. Ligation of the ends of the two strands.

The 3.5-kb RNA transcript, core antigen, and polymerase form core particles, and the polymerase converts the RNA to DNA in the particles as they form in the cytoplasm.

The genome structure and replication of cauliflower mosaic virus (CaMV), the prototype member of the *Caulimovirus* genus, is reminiscent of that of hepadnaviruses, although there are differences between them. The CaMV genome consists of a gapped, circular, double-stranded DNA molecule of about 8 kbp, one strand of which is known as the α-strand and contains a

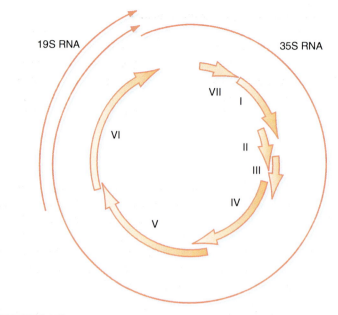

FIGURE 3.23 CaMV genome.
Structure, organization, and proteins encoded by the CaMV genome.

single gap, and a complementary strand, which contains two gaps (Figure 3.23). There are eight genes encoded in this genome, although not all eight products have been detected in infected cells. Replication of the CaMV genome is similar to that of HBV. The first stage is the migration of the gapped virus DNA to the nucleus of the infected cell where it is repaired to form a covalently closed circle. This DNA is transcribed to produce two polyadenylated transcripts, one long (35S) and one shorter (19S). In the cytoplasm, the 19S mRNA is translated to produce a protein that forms large **inclusion bodies** in the cytoplasm of infected cells, and it is in these sites that the second phase of replication occurs. In these replication complexes, some copies of the 35S mRNA are translated while others are reverse transcribed and packaged into **virions** as they form. The differences between reverse transcription of these virus genomes and those of retroviruses are summarized in Table 3.3.

THE VIROME—EVOLUTION AND EPIDEMIOLOGY

Vast number of virus particles populate this planet. Our best estimate is that there are 10^{31} virus particles on Earth. Viruses outnumber bacteria 10 to 1 in most ecosystems, including our bodies. Years ago we abandoned the idea

Table 3.3 Reverse Transcription of Virus Genomes

Features	Caulimoviruses	Hepadnaviruses	Retroviruses
Genome	DNA	DNA	RNA
Primer for (−) strand synthesis	tRNA	Protein	tRNA
Terminal repeats (LTRs)	No	No	Yes
Specific integration of virus genome	No	No	Yes

that viruses are only associated with diseases—we have known that virus infection is much more prevalent than that—indeed, universal—for decades. But as the powerful techniques of molecular biology described in Chapter 1 have matured, we have come to realize how complex and abundant the virus world is. This knowledge has resulted in the development of a new concept in recent years, that of the virome, that is, all the viruses in a given environment, for example, in an ocean, or in the human body.

Epidemiology is concerned with the distribution of disease and the developing strategies to reduce or prevent it. Virus infections present considerable difficulties for this process. Except for epidemics where acute symptoms are obvious, the major evidence of virus infection available to the epidemiologist is the presence of antivirus antibodies in patients. This information frequently provides an incomplete picture, and it is often difficult to assess whether a virus infection occurred recently or at some time in the past. Techniques such as the isolation of viruses in experimental plants or animals are laborious and impossible to apply to large populations. Molecular biology provides sensitive, rapid, and sophisticated techniques to detect and analyze the genetic information stored in virus genomes and has resulted in a new area of investigation: molecular epidemiology, informed by our growing knowledge of the human virome.

Similarities in the coat protein structures of archaeal viruses and those of bacterial and animal viruses suggest that at least some present-day viruses may have a common ancestor that precedes the division into three domains of life over three billion years ago, suggesting that viruses have lineages that can be traced back to near the root of the universal tree of life, the "last universal common ancestor." At least three theories seek to explain the origin of viruses:

- **Regressive evolution:** This theory states that viruses are degenerate life forms that have lost many functions that other organisms possess and have only retained the genetic information essential to their parasitic way of life.

Table 3.4 Orders of Related Virus Families

Order	Families
Caudovirales (tailed bacteriophages)	Myoviridae, Podoviridae, Siphoviridae
Herpesvirales (herpesvirus-like)	Alloherpesviridae, Herpesviridae, Malacoherpesviridae
Mononegavirales (nonsegmented negative-sense RNA viruses)	Bornaviridae, Filoviridae, Paramyxoviridae, Rhabdoviridae
Nidovirales ("nested" viruses, because of their pattern of transcription)	Arteriviridae, Coronaviridae, Roniviridae
Picornavirales (picornavirus-like)	Dicistroviridae, Iflaviridae, Marnaviridae, Picornaviridae, Secoviridae
Tymovirales (tymovirus-like)	Alphaflexiviridae, Betaflexiviridae, Gammaflexiviridae, Tymoviridae

- **Cellular origins:** In this theory, viruses are thought to be subcellular, functional assemblies of macromolecules that have escaped their origins inside cells.
- **Independent entities:** This theory suggests that viruses evolved on a parallel course to cellular organisms from the self-replicating molecules believed to have existed in the primitive, prebiotic RNA world.

While each of these theories has its devotees and this subject provokes fierce disagreements, the fact is that viruses exist, and we are all infected with them. The practical importance of the origin of viruses is that this issue may have implications for virology here and now. Genetic and nucleotide sequence relationships between viruses can reveal the origins not only of individual viruses, but also of whole families and possible "superfamilies." In a number of groups of viruses previously thought to be unrelated, genome sequencing has revealed that functional regions appear to be grouped together in a similar way. The extent to which there is any sequence similarity between these regions in different viruses varies, although clearly, the active sites of enzymes such as virus replicases are strongly conserved. The emphasis in these groups is more on functional and organizational similarities. The original classification scheme for viruses did not recognize a higher level grouping than the family (see Appendix 2), but there are currently seven groups of related virus families equivalent to the orders of formal biological nomenclature (Table 3.4).

Knowledge drawn from taxonomic relationships allows us to predict the properties and behavior of new viruses or to develop drugs based on what is already known about existing viruses. It is believed these shared patterns suggest the descent of present-day viruses from a limited number of primitive ancestors. Although it is tempting to speculate on events that may have

BOX 3.3 WHAT DO ORDERS TELL US ABOUT EVOLUTION?

When the International Committee on Nomenclature of Viruses was created in 1966, we knew hundreds of viruses but little about most of them. This made it difficult to see how they were related to each other. Eventually it was agreed that some viruses were sufficiently similar to allow them to be grouped together as a genus—in the same way that horses (*Equus caballus*) are in the same genus as donkeys (*Equus asinus*). The next step was to group similar genera (plural of genus) together as families. At that point, there was a pause for some years until it was agreed that similar virus families could be grouped into orders, of

which seven have now been recognized. This change happened after enough nucleotide sequence data had been accumulated to make the faint evolutionary relationships between distantly related viruses apparent. Why does it matter? In part because this is a window on the past allowing us to look back millions of years through these genetic fossils, but much more importantly because it points to what viruses are capable of and where they might be going in the future. And that's something we should all worry about.

occurred before the origins of life as it is presently recognized, it would be unwise to discount the pressures that might result in viruses with diverse origins assuming common genetic solutions to common problems of storing, replicating, and expressing genetic information. This is particularly true now that we appreciate the plasticity of virus and cellular genomes and the mobility of genetic information from virus to virus, cell to virus, and virus to cell. There is no reason to believe that virus evolution has stopped, and it is dangerous to do so. The practical consequences of ongoing evolution and the concept of emergent viruses are described in Chapter 7.

SUMMARY

Molecular biology has put much emphasis on the structure and function of the virus genome. At first sight, this tends to emphasize the tremendous diversity of virus genomes. On closer examination, similarities and unifying themes become more apparent. Sequences and structures at the ends of virus genomes are in some ways functionally more significant than the unique coding regions within them. Common patterns of genetic organization seen in virus superfamilies and orders suggest either that many viruses have evolved from common ancestors or that exchange of genetic information between viruses has resulted in common solutions to common problems.

Further Reading

Barr, J.N., Fearns, R., 2010. How RNA viruses maintain their genome integrity. J. Gen. Virol. 91 (6), 1373–1387.

Bieniasz, P.D., 2009. The cell biology of HIV-1 virion genesis. Cell Host Microbe 5 (6), 550–558. Available from: http://dx.doi.org/10.1016/j.chom.2009.05.015.

Delwart, E., 2013. A roadmap to the human virome. PLoS Pathog. 9 (2), e1003146. Available from: http://dx.doi.org/10.1371/journal.ppat.1003146.

Domingo, E., Holland, J.J., 2010. The origin and evolution of viruses. Topley & Wilson's Microbiology and Microbial Infections. John Wiley & Sons.

Forterre, P., Prangishvili, D., 2009. The great billion-year war between ribosome- and capsid-encoding organisms (Cells and Viruses) as the major source of evolutionary novelties. Ann. N. Y. Acad. Sci. 1178, 65−77. Available from: http://dx.doi.org/10.1111/j.1749-6632.2009.04993.x.

Hutchinson, E.C., von Kirchbach, J.C., Gog, J.R., Digard, P., 2010. Genome packaging in influenza A virus. J. Gen. Virol. 91, 313−328. Available from: http://dx.doi.org/10.1099/vir.0.017608-0.

Koonin, E.V., Senkevich, T.G., Dolja, V.V., 2006. The ancient virus world and evolution of cells. Biol. Direct 1 (1), 29.

Legendre, M., Bartoli, J., et al., 2014. Thirty-thousand-year-old distant relative of giant icosahedral DNA viruses with a pandoravirus morphology. Proc. Natl. Acad. Sci. USA 111 (11), 4274−4279.

Nguyen, M., Haenni, A.L., 2003. Expression strategies of ambisense viruses. Virus Res. 93, 141−150, doi:10.1016/S0168-1702(03)00094-7.

Saey, T.H., 2014. Beyond the microbiome: the vast virome: scientists are just beginning to get a handle on the many roles of viruses in the human ecosystem. Sci. News 185 (1), 18−21.

Van Etten, J.L., Lane, L.C., Dunigan, D.D., 2010. DNA Viruses: the really big ones (Giruses). Annu. Rev. Microbiol. 64, 83−99. Available from: http://dx.doi.org/10.1146/annurev.micro.112408.134338.

CHAPTER 4

Replication

Intended Learning Outcomes

On completing this chapter you should be able to:

- Explain the phases of virus replication.
- Describe the key experiments which allowed us to understand what happens during replication.
- Discuss how drugs which block these processes can be used to combat virus infections.

OVERVIEW OF VIRUS REPLICATION

Understanding the details of virus replication is very important. This is not just for academic reasons, but also because this knowledge provides the key to fighting virus infections. The type of cell infected by the virus has a profound effect on the process of replication. For viruses of prokaryotes, replication tends to reflect the relative simplicity of their host cells. For viruses with eukaryotic hosts, processes are frequently more complex. There are many examples of animal viruses undergoing different replicative cycles in different cell types; however, the coding capacity of the virus genome forces all viruses to choose a strategy for replication. This might be one involving heavy reliance on the host cell, in which case the virus genome can be very compact and need only encode the essential information for a few proteins (e.g., parvoviruses). Alternatively, large and complex virus genomes, such as those of poxviruses, encode most of the information necessary for replication, and the virus is only reliant on the cell for the provision of energy and the apparatus for macromolecular synthesis, such as ribosomes (see Chapter 1). Viruses with an RNA lifestyle (i.e., an RNA genome plus messenger RNAs) have no apparent need to enter the nucleus, although during the course of replication a few do. DNA viruses, as might be expected, mostly replicate in the nucleus, where host-cell DNA is replicated and where the biochemical apparatus

Principles of Molecular Virology. DOI: http://dx.doi.org/10.1016/B978-0-12-801946-7.00004-3

necessary for this process is located. However, some viruses with DNA genomes (e.g., poxviruses) have evolved to contain sufficient biochemical capacity to be able to replicate in the cytoplasm, with minimal requirement for host-cell functions. Most of this chapter will examine the process of virus replication and will look at some of the variations on the basic theme.

INVESTIGATION OF VIRUS REPLICATION

Bacteriophages have long been used by virologists as models to understand the biology of viruses. This is particularly true of virus replication. Two very significant experiments that illustrated the fundamental nature of viruses were performed on bacteriophages. The first of these was done by Ellis and Delbruck in 1939 and is usually referred to as the "single-burst" experiment or "one-step growth curve" (Figure 4.1). This was the first experiment to show the three essential phases of virus replication:

- Initiation of infection,
- Replication and expression of the virus genome,
- Release of mature **virions** from the infected cell.

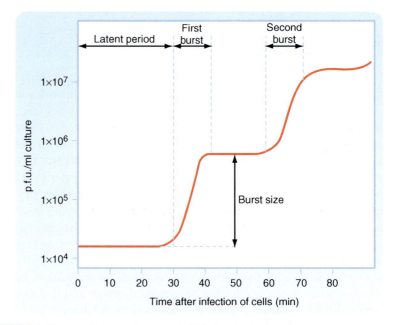

FIGURE 4.1 The one-step growth curve or single-burst experiment.
First performed by Ellis and Delbruck in 1939, this classic experiment illustrates the true nature of virus replication. Details of the experiment are given in the text. Two bursts (crops of phage particles released from cells) are shown in this particular experiment.

In this experiment, bacteriophage particles were added to a culture of rapidly growing bacteria, and after a period of a few minutes, the culture was diluted, effectively preventing further interaction between the **phage** particles and the cells. This simple step is the key to the entire experiment, as it effectively synchronizes the infection of the cells and allows the subsequent phases of replication in a population of individual cells and virus particles to be viewed as if it were a single interaction (in much the same way that molecular cloning of nucleic acids allows analysis of populations of nucleic acid molecules as single species). Repeated samples of the culture were taken at short intervals and analyzed for bacterial cells by plating onto agar plates and for phage particles by plating onto lawns of bacteria. As can be seen in Figure 4.1, there is a stepwise increase in the concentration of phage particles with time, each increase in phage concentration representing one replicative cycle of the virus. However, the data from this experiment can also be analyzed in a different way, by plotting the number of **plaque-forming units** (**p.f.u.**) per bacterial cell against time (Figure 4.2). In this type of assay, a p.f.u.

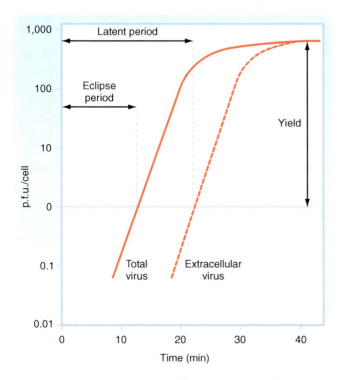

FIGURE 4.2 Analysis of data from a single-burst experiment.
Unlike Figure 4.1, which shows the total number of p.f.u. produced, here the data are plotted as p.f.u./ bacterial cell, reflecting the events occurring in a "typical" infected cell in the population. The phases of replication named on the graph are defined in the text.

can be either a single extracellular virus particle or an infected bacterial cell. These two can be distinguished by disruption of the bacteria with chloroform before plating, which releases any intracellular phage particles, thus providing the total virus count (i.e., intracellular plus extracellular particles).

Several additional features of virus replication are visible from the graph in Figure 4.2. Immediately after dilution of the culture, there is a phase of 10−15 min when no phage particles are detectable; this is known as the eclipse period. This represents a time when virus particles have broken down after penetrating cells, releasing their genomes as a prerequisite to replication. At this stage, they are no longer infectious and therefore cannot be detected by the plaque assay. The latent period is the time before the first new extracellular virus particles appear and is around 20−25 min for many bacteriophages. About 40 min after the cells have been infected, the curves for the total number of virus particles and for extracellular virus merge because the infected cells have lysed and released any intracellular phage particles by this time. The yield (i.e., number) of particles produced per infected cell can be calculated from the overall rise in phage titer.

Following the development of plaque assays for animal viruses in the 1950s, single-burst experiments have been performed for many viruses of eukaryotes, with similar results (Figure 4.3). The major difference between these viruses and bacteriophages is the much longer time interval required for replication, which is measured in terms of hours and, in some cases, days, rather than minutes. This difference reflects the much slower growth rate of eukaryotic cells and, in part, the complexity of virus replication in compartmentalized cells. Biochemical analysis of virus replication in eukaryotic cells has also been used to analyze the levels of virus and cellular protein and nucleic acid synthesis and to examine the intracellular events occurring during synchronized infections (Figure 4.4). The use of various metabolic inhibitors, discussed later in this chapter, also proved to be a valuable tool in such experiments.

The second key experiment on virus replication using bacteriophages was performed by Hershey and Chase in 1952. Bacteriophage T2 was propagated in *Escherichia coli* cells that had been labeled with one of two radioisotopes, either ^{35}S, which is incorporated into sulfur-containing amino acids in proteins, or ^{32}P, which is incorporated into nucleic acids (which do not contain any sulfur) (Figure 4.5). Particles labeled in each of these ways were used to infect bacteria. After a short period to allow attachment to the cells, the mixture was homogenized briefly in a blender which did not destroy the bacterial cells but was sufficiently vigorous to knock the phage coats off the outside of the cells. Analysis of the radioactive content in the cell pellets and

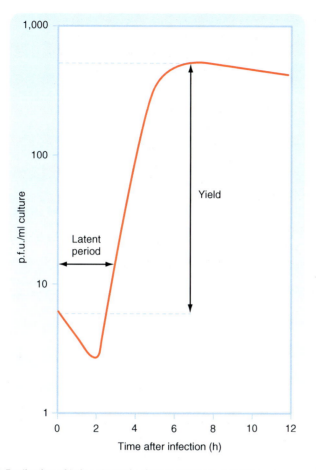

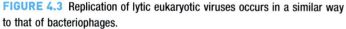

FIGURE 4.3 Replication of lytic eukaryotic viruses occurs in a similar way to that of bacteriophages.

This figure shows a single-burst type of experiment for a picornavirus (e.g., poliovirus). This type of data can only be produced from synchronous infections where a high **multiplicity of infection** is used.

culture supernatant (containing the empty phage coats) showed that most of the radioactivity in the ^{35}S-labeled particles remained in the supernatant, while in the ^{32}P-labeled particles most of the radiolabel had entered the cells. Additionally, they showed that adding deoxyribonuclease (DNase), an enzyme that breaks down DNA, into a suspension containing the labeled bacteriophages did not introduce any ^{32}P into solution. This demonstrated that the phage particles are resistant to the enzyme while intact. In contrast, disrupting the particles by osmotic shock creates a soluble fraction containing most of the ^{32}P and a heavier fraction that contains the ^{35}S protein coat

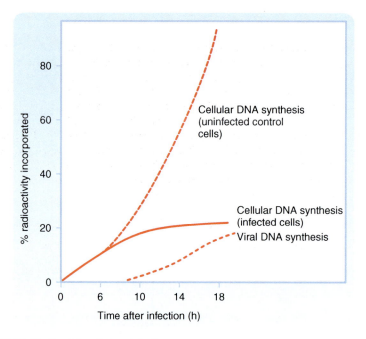

FIGURE 4.4 Biochemistry of virus infection.
This graph shows the rate of cellular and virus DNA synthesis (based on the incorporation of
radiolabeled nucleotides into high-molecular-weight material) in uninfected and virus-infected cells.

of the virus. It was found that the "ghost" particles in the heavy fraction
could still attach to bacteria even though they contained no DNA. This further evidence helped to differentiate the roles of the DNA and protein components of the virus.

This experiment proved that it was the DNA **genome** of the bacteriophage
that entered the cells and initiated the infection rather than any other component (such as proteins). Although it might seem obvious now, at the
time this experiment settled a great controversy over whether a structurally
simple polymer such as a nucleic acid, which was known to contain only
four monomers, was complex enough to carry genetic information. (At the
time, it was generally believed that proteins, which consist of a much more
complex mixture of more than 20 different amino acids, were the carriers
of the genes and that DNA was probably a structural component of cells
and viruses.) Together, these two experiments illustrate the essential processes of virus replication. Virus particles enter susceptible cells and release
their genomic nucleic acids. These are replicated and packaged into virus
particles consisting of newly synthesized virus proteins, which are then
released from the cell.

BOX 4.1 SEEING THE WOOD FOR THE TREES

Henry Ford said "History is bunk"—but he was wrong. Obscure experiments from the 1930s might not seem very interesting, but if you think that, you're making the same mistake as Henry. In theory, it would be very simple to repeat Ellis and Delbruck's experiment in a modern virus research laboratory—except that it's unlikely that this would happen. Bacteriophages don't make people sick (very often—more about that later in the book), so they don't get much attention these days when the only way you can run a laboratory is to get lots of research grants for working on "important" viruses such as HIV. Except

that if you tried to do the Ellis and Delbruck experiment on HIV, you wouldn't be able to, because of the biology of this virus. Although it goes through all the same stages of replication as a bacteriophage, you wouldn't be able to interpret the data you got because of the kinetics. Historically, bacteriophages have easily been the most important "model organisms" in virology, and continue to give us insights into diversity, adaptation, and virulence which are much harder to study in "more advanced" viruses.

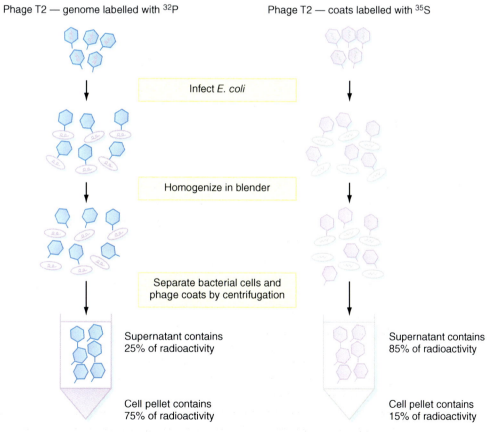

Phage T2 — genome labelled with ^{32}P   Phage T2 — coats labelled with ^{35}S

Infect *E. coli*

Homogenize in blender

Separate bacterial cells and phage coats by centrifugation

Supernatant contains 25% of radioactivity

Cell pellet contains 75% of radioactivity

Supernatant contains 85% of radioactivity

Cell pellet contains 15% of radioactivity

FIGURE 4.5 The Hershey–Chase experiment.

The Hershey–Chase experiment, first performed in 1952, demonstrated that virus genetic information was encoded by nucleic acids and not proteins. Details of the experiment are described in the text.

THE REPLICATION CYCLE

Virus replication can be divided into eight stages, as shown in Figure 4.6. These are purely arbitrary steps, used here for convenience in explaining the replication cycle of a nonexistent "typical" virus. For simplicity, this chapter concentrates on viruses that infect vertebrates. Viruses of bacteria, invertebrates, and plants are mentioned briefly, but the overall objective of this chapter is to illustrate similarities in the pattern of replication of different viruses. Regardless of their hosts, all viruses must undergo each of these stages in some form to successfully complete their replication cycles. Not all the steps described here are detectable as distinct stages for all viruses—often they blur together and appear to occur almost simultaneously. Some of the individual stages have been studied in great detail, and a tremendous amount of information is known about them. Other stages have been much more difficult to study, and considerably less information is available.

Attachment

Because the separate stages of virus replication described here are arbitrary and because complete replication necessarily involves a cycle, it is possible to begin discussion of virus replication at any point. Arguably, it is most logical to consider the first interaction of a virus with a new host cell as the starting point of the cycle. Technically, virus **attachment** consists of specific binding of a **virus-attachment protein** (or "antireceptor") to a cellular **receptor** molecule. Many examples of virus receptors are now known (see Figure 4.7 and the Further Reading at the end of this chapter). The target receptor molecules on cell surfaces may be proteins (usually glycoproteins) or the carbohydrate structures present on glycoproteins or glycolipids. The former are usually specific receptors in that a virus may use a particular protein as a receptor.

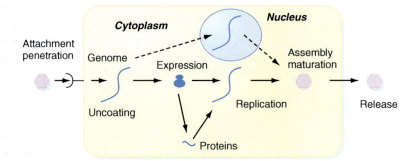

FIGURE 4.6 A generalized scheme for virus replication.
This diagram shows an outline of the steps which occur during replication of a "typical" virus which infects eukaryotic cells. See the text for more details.

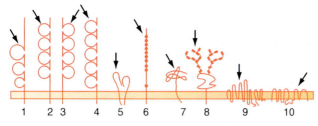

FIGURE 4.7 Virus receptors.
The arrows in this figure indicate approximate virus-attachment sites. (1) PVR. (2) CD4: HIV.
(3) Carcinoembryonic antigen(s): MHV (coronavirus). (4) ICAM-1: most rhinoviruses. (Note that 1—4
are all immunoglobulin superfamily molecules.) (5) VLA-2 integrin: ECHO viruses. (6) LDL receptor: some
rhinoviruses. (7) Aminopeptidase N: coronaviruses. (8) Sialic acid (on glycoprotein): influenza, reoviruses,
rotaviruses. (9) Cationic amino acid transporter: murine leukemia virus. (10) Sodium-dependent
phosphate transporter: Gibbon ape leukemia virus.

Carbohydrate groups are usually less specific because the same configuration
of sugar side-chains may occur on many different glycosylated membrane-
bound molecules. Some complex viruses (e.g., poxviruses, herpesviruses) use
more than one receptor and therefore have alternative routes of uptake into
cells. Virus receptors fall into many different classes (e.g., immunoglobulin-
like superfamily molecules, membrane-associated receptors, and transmembrane (TM) transporters and channels). The one factor that unifies all virus
receptors is that they did not evolve and are not manufactured by cells to
allow viruses to enter cells—rather, viruses have subverted molecules
required for normal cellular functions.

Plant viruses face special problems initiating infection. The outer surfaces of
plants are composed of protective layers of waxes and pectin, but more significantly, each cell is surrounded by a thick wall of cellulose overlying the
cytoplasmic membrane. To date, no plant virus is known to use a specific cellular receptor of the type that animal and bacterial viruses use to attach to
cells. Instead, plant viruses rely on a breach of the integrity of a cell wall to
introduce a virus particle directly into a host cell. This is achieved either by
the vector associated with transmission of the virus or simply by mechanical
damage to cells. After replication in an initial cell, the lack of receptors poses
further problems for plant viruses in recruiting new cells to the infection.
These are discussed in Chapter 6.

Some of the best understood examples of virus—receptor interactions are
from the *Picornaviridae*. The virus—receptor interaction in picornaviruses has
been studied intensively from the viewpoint of both the structural features of
the virus responsible for receptor binding and those of the receptor molecule.
The major human rhinovirus (HRV) receptor molecule, ICAM-1 (intercellular

adhesion molecule 1 or CD54), is an adhesion molecule whose normal function is to bind cells to adjacent substrates. Structurally, ICAM-1 is similar to an immunoglobulin molecule, with constant (C) and variable (V) domains homologous to those of antibodies and is regarded as a member of the immunoglobulin superfamily of proteins (Figure 4.7). Similarly, the poliovirus receptor, PVR, or CD155, is an integral membrane protein that is also a member of this family, with one variable and two constant domains which is involved in establishment of intercellular junctions between epithelial cells.

Since the structure of a number of picornavirus capsids is known at a resolution of a few angstroms (Chapter 2), it has been possible to determine the features of the virus responsible for receptor binding. In HRVs, there is a deep cleft known as the "canyon" in the surface of each triangular face of the icosahedral capsid, which is formed by the flanking monomers, VP1, VP2, and VP3 (Figure 4.8). Biochemical evidence from a class of inhibitory drugs that block attachment of HRV particles to cells indicates that the interaction between ICAM-1 and the virus particle occurs on the floor of this canyon. Unlike other areas of the virus surface, the amino acid residues forming the internal surfaces of the canyon are relatively invariant. It was suggested that these regions are protected from antigenic pressure because the antibody molecules are too large to fit into the cleft. This is important because radical changes here, although allowing the virus to escape an immune response, would disrupt receptor binding. Subsequently, it has been found that the binding site of the receptor extends well over the edges of the canyon, and the binding sites of neutralizing antibodies extend over the rims of the canyon. Nevertheless, the residues most significant for the binding site of the receptor and for neutralizing antibodies are separated from each other. In polioviruses, there is a similar canyon that runs around each fivefold vertex of the capsid. The highly variant regions of the capsid to which antibodies

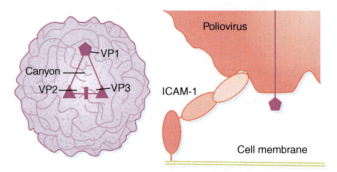

FIGURE 4.8 Rhinovirus receptor binding.
Rhinovirus particles have a deep surface cleft, known as the "canyon," between the three monomers (VP1, 2, and 3) making up each face of the particle.

bind are located on the "peaks" on either side of this trough, which is again too narrow to allow antibody binding to the residues at its base. The invariant residues at the sides of the trough interact with the receptor.

Even within the *Picornaviridae* there is considerable variation in receptor usage. Although 90 serotypes of HRV use ICAM-1 as their receptor, some 10 serotypes use proteins related to the low-density lipoprotein (LDL) receptor. Encephalomyocarditis virus has been reported to use the immunoglobulin molecule vascular cell adhesion factor or glycophorin A. Several picornaviruses use other integrins as receptors: some enteric cytopathic human orphan (ECHO) viruses use VLA-2 or fibronectin, and foot-and-mouth disease viruses have been reported to use an unidentified integrin-like molecule. Other ECHO viruses use complement decay-accelerating factor (DAF, CD55), a molecule involved in complement regulation. This list is given to illustrate that even within one structurally closely related family of viruses, there is considerable variation in the receptor structures used.

Another well-studied example of a virus–receptor interaction is that of influenza virus. The influenza virus hemagglutinin protein forms one of the two types of glycoprotein spikes on the surface of the particles (see Chapter 2), the other type being formed by the neuraminidase protein. Each hemagglutinin spike is composed of a trimer of three molecules, while the neuraminidase spike consists of a tetramer (Figure 4.9). The hemagglutinin spikes are

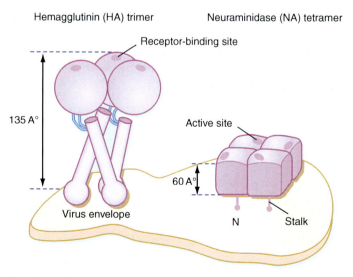

FIGURE 4.9 Influenza virus glycoprotein spikes.
The glycoproteins spikes on the surface of influenza virus (and many other enveloped viruses) are multimers consisting of three copies of the hemagglutinin protein ("trimer") and four copies of the neuraminidase protein ("tetramer").

responsible for binding the influenza virus receptor, which is sialic acid (*N*-acetyl neuraminic acid), a sugar group commonly found on a variety of glycosylated molecules. As a result, little cell-type specificity is imposed by this receptor interaction so influenza viruses bind to a wide variety of different cell types (e.g., causing hemagglutination of red blood cells) in addition to the cells in which productive infection occurs.

The neuraminidase molecule of influenza virus and paramyxoviruses illustrates another feature of this stage of virus replication. Attachment to cellular receptors is in most cases a reversible process—if penetration of the cells does not ensue, the virus can elute from the cell surface. Some viruses have specific mechanisms for detachment, and the neuraminidase protein is one of these. Neuraminidase is an esterase that cleaves sialic acid from sugar side-chains. This is particularly important for influenza. Because the receptor molecule is so widely distributed, the virus tends to bind inappropriately to a variety of cells and even cell debris; however, elution from the cell surface after receptor binding has occurred often leads to changes in the virus (e.g., loss or structural alteration of virus-attachment protein) which decrease or eliminate the possibility of subsequent attachment to other cells. Thus, in the case of influenza, cleavage of sialic acid residues by neuraminidase leaves these groups bound to the active site of the hemagglutinin, preventing that particular molecule from binding to another receptor.

BOX 4.2 WHY DO THESE OBSCURE DETAILS MATTER?

I've spent quite a long time in this chapter describing the interactions of certain viruses with their receptors. If you look at the research literature about virus receptors, you'll find it's huge. Why all the fuss? It's because this first interaction of a virus particle with a host cell is in some ways the most important step in replication—it goes a long way to determining what happens in the rest of the process. For one thing, if a cell has no receptors for a virus, it doesn't get infected. So tropism, the ability to infect a particular cell type, is largely controlled by receptor interactions (and only occasionally by events inside the cell). Going on from there, small changes can have big effects, so this process is important to understand in detail. At present, the H5N1 type of influenza virus can infect humans and when it does, it's likely to kill them, but it really struggles to do this because at the moment, it's really a bird (avian) virus. With a very small change in the receptor usage, H5N1 could become a deadly human virus. In addition, when you understand these processes, you can use them against the virus. We've had anti-influenza drugs for decades, but they weren't very good. Modern influenza drugs such as Tamiflu and Relenza inhibit the neuraminidase protein involved in receptor interactions (although in release from the cell rather than uptake). If H5N1 ever does make the jump to being a human virus, we're going to need these drugs to stay alive.

In most cases, the expression (or absence) of receptors on the surface of cells largely determines the tropism of a virus (i.e., the type of host cell in which it is able to replicate). In some cases, intracellular blocks at later stages of replication are responsible for determining the range of cell types in which a virus can carry out a productive infection, but this is not common. Therefore, this initial stage of replication and the very first interaction between the virus and the host cell has a major influence on virus pathogenesis and in determining the course of a virus infection. In some cases, interactions with more than one protein are required for virus entry. These are not examples of alternative receptor use, as neither protein alone is a functional receptor—both are required to act together. An example is the process by which adenoviruses enter cells. This requires a two-stage process involving an initial interaction of the virion fiber protein with a range of cellular receptors, which include the major histocompatibility complex class I (MHC-I) molecule and the coxsackievirus–adenovirus receptor. Another virion protein, the penton base, then binds to the integrin family of cell-surface proteins, allowing internalization of the particle by receptor-mediated endocytosis. Most cells express primary receptors for the adenovirus fiber coat protein; however, the internalization step is more selective, giving rise to a degree of cell selection.

A similar observation has been made with human immunodeficiency virus (HIV). The primary receptor for HIV is the helper T-cell differentiation antigen, CD4. Transfection of human cells that do not normally express CD4 (such as epithelial cells) with recombinant CD4-expression constructs makes them permissive for HIV infection; however, transfection of rodent cells with human CD4-expression vectors does not permit productive HIV infection—something else is missing from the mouse cells. If HIV provirus DNA is inserted into rodent cells by transfection, virus is produced, showing that there is no intracellular block to infection. So there must be one or more accessory factors in addition to CD4 that are required to form a functional HIV receptor. These are a family of proteins known as β-chemokine receptors. Several members of this family have been shown to play a role in the entry of HIV into cells, and their distribution may be the primary control for the tropism of HIV for different cell types (lymphocytes, macrophages, etc.). Furthermore, there is evidence, in at least some cell types, that HIV infection is not blocked by competing soluble CD4, indicating that in these cells a completely different receptor strategy may be being used. Several candidate molecules have been put forward to fill this role (e.g., galactosylceramide and various other candidate proteins). However, if any or all of these do allow HIV to infect a range of CD4-negative cells, this process is much less efficient than the interaction of the virus with its major receptor complex.

In some cases, specific receptor binding can be side-stepped by nonspecific or inappropriate interactions between virus particles and cells. It is possible that virus particles can be "accidentally" taken up by cells via processes such as pinocytosis or phagocytosis. However, in the absence of some form of physical interaction that holds the virus particle in close association with the cell surface, the frequency with which these accidental events happen is very low. On occasion, antibody-coated virus particles binding to Fc receptor molecules on the surface of monocytes and other blood cells can result in virus uptake. This phenomenon has been shown to occur in a number of cases where antibody-dependent enhancement of virus uptake results in unexpected findings. For example, the presence of antivirus antibodies can occasionally result in increased virus uptake by cells and increased pathogenicity rather than virus neutralization, as would normally be expected. It has been suggested that this mechanism may also be important in the uptake of HIV by macrophages and monocytes and that this is a factor in the pathogenesis of acquired immune deficiency syndrome (AIDS).

Penetration

Penetration of the target cell normally occurs a very short time after attachment of the virus to its receptor in the cell membrane. Unlike attachment, cell penetration is generally an energy-dependent process—that is, the cell must be metabolically active for this to occur. Three main mechanisms are involved:

1. Translocation of the entire virus particle across the cytoplasmic membrane of the cell (Figure 4.10). This process is relatively rare among viruses and is poorly understood. It must be mediated by proteins in the virus capsid and specific membrane receptors.
2. Endocytosis of the virus into intracellular vacuoles (Figure 4.11). This is probably the most common mechanism of virus entry into cells. It does not require any specific virus proteins (other than those already utilized for receptor binding) but relies on the normal formation and internalization of coated pits at the cell membrane. Receptor-mediated endocytosis is an efficient process for taking up and concentrating extracellular macromolecules.
3. Fusion of the virus envelope (so this is only applicable to enveloped viruses) with the cell membrane, either directly at the cell surface or following endocytosis in a cytoplasmic vesicle (Figure 4.12), which requires the presence of a specific fusion protein in the virus envelope—for example, influenza hemagglutinin or retrovirus TM glycoproteins. These proteins promote the joining of the cellular and virus membranes which results in the nucleocapsid being deposited directly in the cytoplasm. There are two types of virus-driven membrane fusion: one pH dependent, the other pH independent.

The process of endocytosis is almost universal in animal cells and deserves further consideration (Figure 4.11). The formation of coated pits results in the engulfment of a membrane-bounded vesicle by the cytoplasm of the cell. The lifetime of these initial coated vesicles is very short. Within seconds,

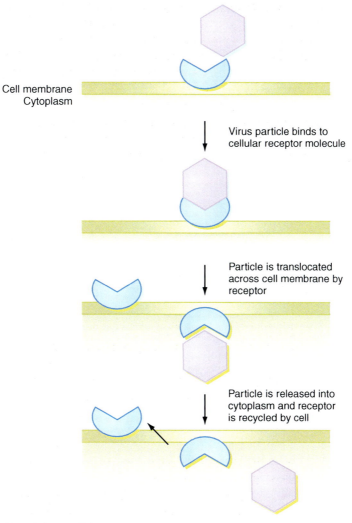

Cell membrane
Cytoplasm

Virus particle binds to
cellular receptor molecule

Particle is translocated
across cell membrane by
receptor

Particle is released into
cytoplasm and receptor
is recycled by cell

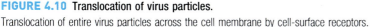

FIGURE 4.10 Translocation of virus particles.
Translocation of entire virus particles across the cell membrane by cell-surface receptors.

most fuse with endosomes, releasing their contents into these larger vesicles. At this point, any virus contained within these structures is still cut off from the cytoplasm by a lipid bilayer and therefore has not strictly entered the cell. Moreover, as endosomes fuse with lysosomes, the environment inside these vessels becomes increasingly hostile as the pH falls, while the concentration of degradative enzymes rises. This means that the virus particle must leave the vesicle and enter the cytoplasm before it is degraded. There are a number of mechanisms by which this can occur, including membrane **fusion**

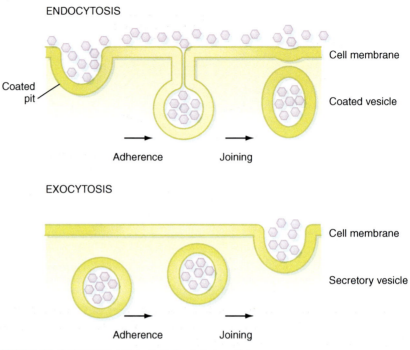

FIGURE 4.11 **Endocytosis and exocytosis of virus particles.**
The processes of endocytosis and exocytosis are involved in both the take up and release of enveloped virus particles from host cells. Viruses modify these normal cellular processes by encoding proteins which promote endocytosis (e.g., virus-attachment proteins and fusion proteins) and release from the cell surface via exocytosis (e.g., the influenza neuraminidase protein).

and rescue by transcytosis. The release of virus particles from endosomes and their passage into the cytoplasm is intimately connected with (and often impossible to separate from) the process of uncoating.

Uncoating

Uncoating is a general term for the events that occur after penetration, in which the virus capsid is completely or partially removed and the virus genome is exposed, usually in the form of a nucleoprotein complex. Unfortunately this is one of the stages of virus replication that has been least studied and is relatively poorly understood. In one sense, the removal of a virus envelope that occurs during membrane fusion is part of the uncoating process. Fusion between virus envelopes and endosomal membranes is driven by virus fusion proteins. These are usually activated by the uncloaking of a previously hidden fusion domain as a result of conformational changes in the protein induced by the low pH inside the vesicle, although in some

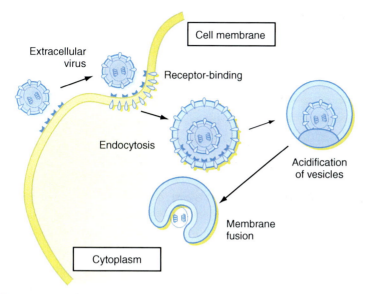

Cell membrane

Extracellular
virus

Receptor-binding

Endocytosis

Acidification
of vesicles

Membrane
fusion

Cytoplasm

FIGURE 4.12 Virus-induced membrane fusion.
This uptake process is dependent on the presence of a specific fusion protein on the surface of the virus which, under particular circumstances (e.g., acidification of the virus-containing vesicle), becomes activated, inducing fusion of the vesicle membrane and the virus envelope.

cases the fusion activity is triggered directly by receptor binding. The initial events in uncoating may occur inside endosomes, being triggered by the change in pH as the endosome is acidified, or directly in the cytoplasm. Proteins which form ion channels, or cations such as chloroquine and ammonium chloride, can be used to block the acidification of these vesicles and to determine whether events are occurring following the acidification of endosomes (e.g., pH-dependent membrane fusion) or directly at the cell surface or in the cytoplasm (e.g., pH-independent membrane fusion). Endocytosis is potentially dangerous for viruses, because if they remain in the vesicle too long they will be irreversibly damaged by acidification or lysosomal enzymes. Some viruses can control this process—for example, the influenza virus M2 protein is a membrane channel that allows entry of hydrogen ions into the nucleocapsid, facilitating uncoating. The M2 protein is multifunctional and also has a role in influenza virus maturation.

In picornaviruses, penetration of the cytoplasm by exit of virus from endosomes is tightly linked to uncoating (Figure 4.13). The acidic environment of the endosome causes a conformational change in the capsid which reveals hydrophobic domains not present on the surface of mature virus particles. The interaction of these hydrophobic patches with the endosomal membrane is believed to form pores through which the genome passes into the cytoplasm.

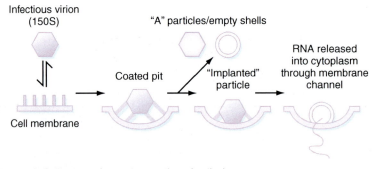

Infectious virion
(150S)

"A" particles/empty shells

RNA released
into cytoplasm
through membrane
channel

Coated pit

"Implanted"
particle

Cell membrane

FIGURE 4.13 Cell penetration and uncoating of polioviruses.
Following receptor binding, poliovirus particles are taken up by host cells in vesicles which interact with the cytoskeleton. This is an active, energy-dependent process.

The product of uncoating depends on the structure of the virus nucleocapsid. In some cases, it might be relatively simple (e.g., picornaviruses have a small basic protein of approximately 23 amino acids [VPg] covalently attached to the 5′ end of the vRNA genome), or highly complex (e.g., retrovirus cores are highly ordered nucleoprotein complexes that contain, in addition to the diploid RNA genome, the reverse transcriptase enzyme responsible for converting the virus RNA genome into the DNA **provirus**). The structure and chemistry of the nucleocapsid determines the subsequent steps in replication. As discussed in Chapter 3, reverse transcription can only occur inside an ordered retrovirus core particle and cannot proceed with the components of the reaction free in solution. Herpesvirus, adenovirus, and polyomavirus **capsids** undergo structural changes following **penetration**, but overall remain largely intact. These capsids contain sequences that are responsible for attachment to the cytoskeleton, and this interaction allows the transport of the entire capsid to the nucleus. It is at the nuclear pores that uncoating occurs and the **nucleocapsid** passes into the nucleus. In reoviruses and poxviruses, complete uncoating does not occur, and many of the reactions of genome replication are catalyzed by virus-encoded enzymes inside cytoplasmic particles which still resemble the mature **virions**.

Genome Replication and Gene Expression

The replication strategy of any virus depends on the nature of its genetic material. In this respect, all viruses can be divided into seven groups. Such a scheme was first proposed by David Baltimore in 1971. Originally, this classification included only six groups, but it has since been extended to include the scheme of **genome** replication used by the hepadnaviruses and caulimoviruses. For viruses with RNA genomes in particular, genome replication and the expression of genetic information are inextricably linked, therefore both

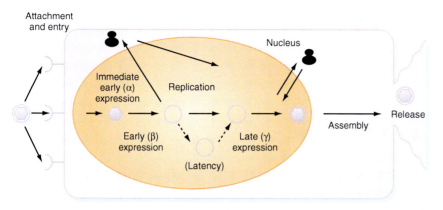

Attachment and entry

FIGURE 4.14 Schematic representation of the replication of class I viruses.
Details of the events that occur for genomes of this type are given in the text.

of these criteria are taken into account in the scheme below. The control of gene expression determines the overall course of a virus infection (acute, chronic, persistent, or latent), and such is the emphasis placed on gene expression by molecular biologists that this subject is discussed in detail in Chapter 5. A schematic overview of the major events during replication of the different virus genomes is shown in Figure 4.14, and a complete list of all the families that constitute each class is given in Appendix 2.

- **Class I:** Double-stranded DNA.
  This class can be subdivided into two further groups: (a) replication is exclusively nuclear (Figure 4.14)—replication of these viruses is relatively dependent on cellular factors; (b) replication occurs in cytoplasm (e.g., the *Poxviridae*), in which case the viruses have evolved (or acquired) all the necessary factors for transcription and replication of their genomes and are therefore largely independent of the cellular machinery.
- **Class II:** Single-stranded DNA (Figure 4.15).
  Replication occurs in the nucleus, involving the formation of a double-stranded intermediate which serves as a template for the synthesis of single-stranded progeny DNA.
- **Class III:** Double-stranded RNA (Figure 4.16).
  These viruses have segmented genomes. Each segment is transcribed separately to produce individual monocistronic mRNAs.
- **Class IV:** Single-stranded (+)sense RNA.
  These can be subdivided into two groups: (a) Viruses with polycistronic mRNA (Figure 4.17)—as with all the viruses in this class, the genome RNA forms the mRNA and is translated to form a polyprotein product, which is subsequently cleaved to form the mature proteins; (b) Viruses with complex transcription, for which two rounds

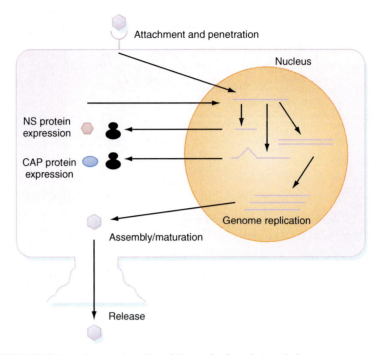

FIGURE 4.15 Schematic representation of the replication of class II viruses.
Details of the events that occur for genomes of this type are given in the text.

of translation (e.g., togavirus) or subgenomic RNAs (e.g., tobamovirus) are necessary to produce the genomic RNA.

- **Class V:** Single-stranded (−)sense RNA.
  As discussed in Chapters 3 and will be discussed in Chapter 5, the genomes of these viruses can be divided into two types: (a) Nonsegmented genomes (order *Mononegvirales*) (Figure 4.18), for which the first step in replication is transcription of the (−)sense RNA genome by the virion RNA-dependent RNA polymerase to produce monocistronic mRNAs, which also serve as the template for subsequent genome replication (note: some of these viruses also have an ambisense organization); (b) Segmented genomes (*Orthomyxoviridae*), for which replication occurs in the nucleus, with monocistronic mRNAs for each of the virus genes produced by the virus transcriptase from the full-length virus genome (see Chapter 5).
- **Class VI:** Single-stranded (+)sense RNA with DNA intermediate (Figure 4.19).
  Retrovirus genomes are (+)sense RNA but unique in that they are diploid and they do not serve directly as mRNA, but as a template for reverse transcription into DNA (see Chapter 3).

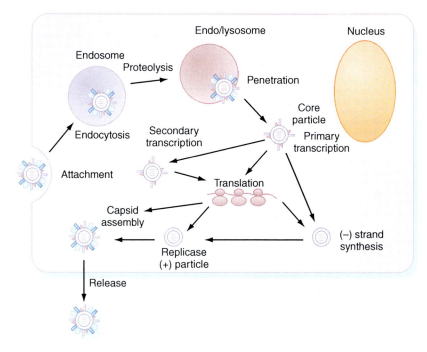

FIGURE 4.16 Schematic representation of the replication of class III viruses.
Details of the events that occur for genomes of this type are given in the text.

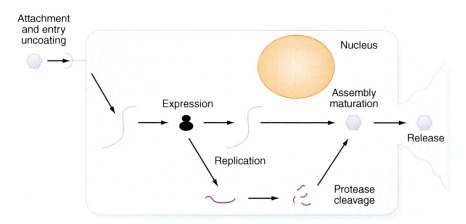

FIGURE 4.17 Schematic representation of the replication of class IV viruses.
Details of the events that occur for genomes of this type are given in the text.

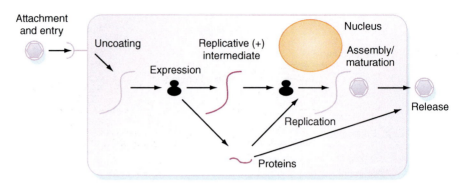

FIGURE 4.18 Schematic representation of the replication of class V viruses.
Details of the events that occur for genomes of this type are given in the text.

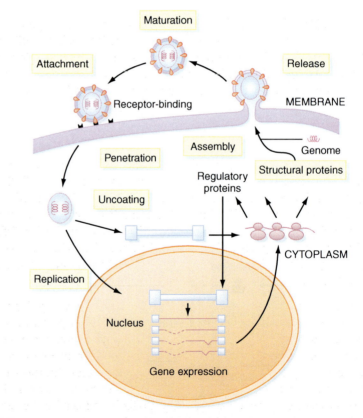

FIGURE 4.19 Schematic representation of the replication of class VI viruses.
Details of the events that occur for genomes of this type are given in the text.

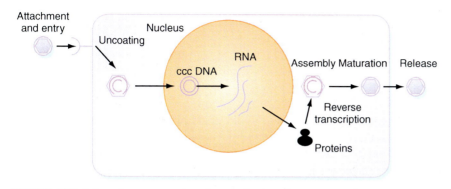

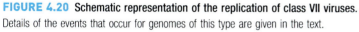

FIGURE 4.20 Schematic representation of the replication of class VII viruses. Details of the events that occur for genomes of this type are given in the text.

- **Class VII:** Double-stranded DNA with RNA intermediate (Figure 4.20). This group of viruses also relies on reverse transcription, but, unlike the retroviruses (class VI), this process occurs inside the virus particle during maturation. On infection of a new cell, the first event to occur is repair of the gapped genome, followed by transcription (see Chapter 3).

Assembly

The **assembly** process involves the collection of all the components necessary for the formation of the mature **virion** at a particular site in the cell. During assembly, the basic structure of the virus particle is formed. The site of assembly depends on the site of replication within the cell and on the mechanism by which the virus is eventually released from the cell and varies for different viruses. For example, in picornaviruses, poxviruses, and reoviruses, assembly occurs in the cytoplasm; in adenoviruses, polyomaviruses, and parvoviruses, it occurs in the nucleus.

Lipid rafts are membrane microdomains enriched with glycosphingolipids (or glycolipids), cholesterol, and a specific set of associated proteins. A high level of saturated hydrocarbon chains in sphingolipids allows cholesterol to be tightly interleaved in these rafts. The lipids in these domains differ from other membrane lipids in having relatively limited lateral diffusion in the membrane, and they can also be physically separated by density centrifugation in the presence of some detergents. Lipid rafts have been implicated in a variety of cellular functions, such as apical sorting of proteins and signal transduction, but they are also used by viruses as platforms for cell entry (e.g., HIV, SV40, and rotavirus), and as sites for particle assembly, budding, and release from the cell membrane (e.g., influenza virus, HIV, measles virus, and rotavirus).

As with the early stages of replication, it is not always possible to identify the assembly, **maturation**, and **release** of virus particles as distinct and separate phases. The site of assembly has a profound influence on all these processes. In the majority of cases, cellular membranes are used to anchor virus proteins, and this initiates the process of assembly. In spite of considerable study, the control of virus assembly is generally not well understood. In general, it is thought that rising intracellular levels of virus proteins and **genome** molecules reach a critical concentration and that this triggers the process. Many viruses achieve high levels of newly synthesized structural components by concentrating these into subcellular compartments, visible in light microscopes, which are known as **inclusion bodies**. These are a common feature of the late stages of infection of cells by many different viruses. The size and location of inclusion bodies in infected cells is often highly characteristic of particular viruses, for example, rabies virus infection results in large perinuclear "Negri bodies," first observed using an optical microscope by Adelchi Negri in 1903. Alternatively, local concentrations of virus structural components can be boosted by lateral interactions between membrane-associated proteins. This mechanism is particularly important in **enveloped** viruses released from the cell by **budding**.

As discussed in Chapter 2, the formation of virus particles may be a relatively simple process which is driven only by interactions between the subunits of the **capsid** and controlled by the rules of symmetry. In other cases, assembly is a highly complex, multistep process involving not only virus structural proteins but also virus-encoded and cellular scaffolding proteins that act as templates to guide the assembly of **virions**. The encapsidation of the virus genome may occur either early in the assembly of the particle (e.g., many viruses with **helical** symmetry are nucleated on the genome) or at a late stage, when the genome is stuffed into an almost completed protein shell.

Maturation

Maturation is the stage of the replication cycle at which the virus becomes infectious. This process usually involves structural changes in the virus particle that may result from specific cleavages of **capsid** proteins to form the mature products or conformational changes which occur in proteins during assembly. Such events frequently lead to substantial structural changes in the capsid that may be detectable by measures such as differences in the antigenicity of incomplete and mature virus particles, which in some cases (e.g., picornaviruses) alters radically. Alternatively, internal structural alterations—for example, the condensation of nucleoproteins with the virus **genome**—often result in such changes. As already stated, for some viruses assembly and maturation occur inside the cell and are inseparable, whereas for others maturation events may occur only after release of the virus particle from the cell. In all cases, the process of maturation prepares the particle for the infection of subsequent cells.

Virus-encoded proteases are frequently involved in maturation, although cellular enzymes or a mixture of virus and cellular enzymes are used in some cases. Clearly there is a danger in relying on cellular proteolytic enzymes in that their relative lack of substrate specificity could easily completely degrade the capsid proteins. In contrast, virus-encoded proteases are usually highly specific for particular amino acid sequences and structures, frequently only cutting one particular peptide bond in a large and complex virus capsid. Moreover, they are often further controlled by being packaged into virus particles during assembly and are only activated when brought into close contact with their target sequence by the conformation of the capsid (e.g., by being placed in a local hydrophobic environment or by changes of pH or metal ion concentration inside the capsid).

Retrovirus proteases are good examples of enzymes involved in maturation which are under this tight control. The retrovirus core particle is composed of proteins from the *gag* gene, and the protease is packaged into the core before its release from the cell on budding. At some stage of the budding process (the exact timing varies for different retroviruses) the protease cleaves the *gag* protein precursors into the mature products—the capsid, nucleocapsid, and matrix proteins of the mature virus particle (Figure 4.21). Not all protease cleavage events involved in maturation are this tightly regulated. Native influenza virus hemagglutinin undergoes posttranslational modification (glycosylation in the Golgi apparatus) and at this stage exhibits receptor-binding activity. However, the protein must be cleaved into two fragments (HA_1 and HA_2) to be able to produce membrane fusion during infection. Cellular trypsin-like enzymes are responsible for this process, which occurs in secretory vesicles as the virus buds into them prior to release at the cell surface. Amantadine and rimantadine are two drugs that are active against influenza A viruses (Chapter 6). The action of these closely related compounds is complex, but they block cellular membrane ion channels. The target for both drugs is the influenza matrix protein (M2), but resistance to the drug may also map to the hemagglutinin gene. The replication of some strains of influenza virus is inhibited at the penetration stage and that of others at maturation. The biphasic action of these drugs results from the inability of drug-treated cells to lower the pH of the endosomal compartment (a function normally controlled by the M2 gene product), and hence to cleave hemagglutinin during maturation. Similarly, retrovirus envelope glycoproteins require cleavage into the surface (SU) and TM proteins for

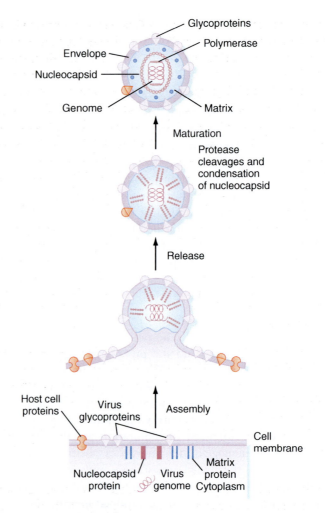

FIGURE 4.21 Virus release by budding.
Budding is the process by which enveloped virus particles acquire their membranes and associated proteins, as well as how they are released from the host cell.

activity. This process is also carried out by cellular enzymes but is in general poorly understood, but it is a target for inhibitors which may act as antiviral drugs.

Release

As described earlier, plant viruses face particular difficulties caused by the structure of plant cell walls when it comes to leaving cells and infecting others. In response, they have evolved particular strategies to overcome this problem which are discussed in detail in Chapter 6. All other viruses escape the cell by one of two mechanisms. For lytic viruses (such as most nonenveloped viruses), release is a simple process—the infected cell breaks open and releases the virus. Enveloped viruses acquire their lipid membrane as the virus buds out of the cell through the cell membrane or into an intracellular vesicle prior to subsequent release. Virion envelope proteins are picked up during this process as the virus particle is extruded. This process is known as budding. Release of virus particles in this way may be highly damaging to the cell (e.g., paramyxoviruses, rhabdoviruses, and togaviruses), or in other cases, appear not to be (e.g., retroviruses), but in either case the process is controlled by the virus—the physical interaction of the capsid proteins on the inner surface of the cell membrane forces the particle out through the membrane (Figure 4.15). As mentioned earlier, assembly, maturation, and release are usually simultaneous processes for virus particles formed by budding. The type of membrane from which the virus buds depends on the virus concerned. In most cases, budding involves cytoplasmic membranes (retroviruses, togaviruses, orthomyxoviruses, paramyxoviruses, bunyaviruses, coronaviruses, rhabdoviruses, and hepadnaviruses) but in some cases can involve the nuclear membrane (herpesviruses).

In a few cases, notably in human retroviruses such as HIV and human T-cell leukemia virus, viruses prefer direct cell-to-cell spread rather than release into the external environment and reuptake by another cell. This process requires intimate contact between cells and can occur at tight junctions between cells or in neurological synapses. These structures have been subverted by human retroviruses which engineer a novel structure in infected cells known as a virological synapse to promote more efficient spread within the host organism.

The release of mature virus particles from susceptible host cells by budding presents a problem in that these particles are designed to enter, rather than leave, cells. How do these particles manage to leave the cell surface? The details are not known but there are clues as to how the process is achieved. Certain virus envelope proteins are involved in the release phase of replication as well as in the initiating steps. A good example of this is the neuraminidase protein of influenza virus. In addition to being able to reverse the attachment of virus particles to cells via hemagglutinin, neuraminidase is also believed to be important in preventing the aggregation of influenza virus particles and may well have a role in virus release. This process is targeted by newer drugs such as oseltamivir (trade name "Tamiflu") and zanamivir ("Relenza") (Chapter 6). In recent years, a group of proteins known as viroporins has been discovered in a range of different viruses. These are small hydrophobic proteins that modify the permeability of cellular membranes and customize host cells for efficient virus propagation, for example, promoting the release of viral particles from infected cells. These proteins are usually not essential for the replication of viruses, but their presence often enhances virus growth.

In addition to using specific proteins, enveloped viruses that bud from cells (see Chapter 2) have also solved the problem of release by the careful timing of the assembly—maturation—release pathway. Although it may not be possible to separate these stages by means of biochemical analysis, this does not mean that careful spatial separation of these processes has not evolved as a means to solve this problem. Similarly, although we may not understand all the subtleties of the many conformation changes that occur in virus **capsids** and envelopes during these late stages of replication, virus replication clearly works, despite our lack of knowledge.

BOX 4.3 WORLDS WITHIN WORLDS

We think of eukaryotic cells as compartmentalized into nucleus and cytoplasm, but the true situation is more complicated than that. There are other biochemical rather than physical compartments within a cell. One is the lipid/aqueous division. Proteins with hydrophobic (water-fearing) domains don't like to be in a soluble form within the cytoplasm. They only start to act when they're in the natural environment of a membrane. But it's not that simple. There are different domains within membranes where different processes occur. Viruses have used these "lipid rafts" for particular functions, such as entering or leaving the cell, and forming tiny factories where new particles are assembled. And then there's time. The process of virus replication doesn't happen in a random order—they are carefully sequenced to optimize the process. This control is directed by the biochemistry of the components involved, which may only start to function as their concentration within an infected cell reaches a critical level. And all of this goes on within the minute world of an infected cell, too small to see with the eye alone, or even the most powerful microscope.

SUMMARY

In general terms, virus replication involves three broad stages carried out by all types of virus: the initiation of infection, replication, and expression of the **genome**, and, finally, **release** of mature **virions** from the infected cell. At a detailed level, there are many differences in the replication processes of different viruses which are imposed by the biology of the host cell and the nature of the virus genome. Nevertheless, it is possible to derive an overview of virus replication and the common stages which, in one form or another, are followed by all viruses.

Further Reading

Bilkova, E., Forstova, J., Abrahamyan, L., 2014. Coat as a dagger: the use of capsid proteins to perforate membranes during non-enveloped DNA viruses trafficking. Viruses 6 (7), 2899–2937.

Ellis, E.L., Delbruck, M., 1939. The growth of bacteriophage. J. Gen. Physiol. 22, 365–384.

Giorda, K.M., Hebert, D.N., 2013. Viroporins customize host cells for efficient viral propagation. DNA Cell Biol. 32 (10), 557–564.

Grove, J., Marsh, M., 2011. The cell biology of receptor-mediated virus entry. J. Cell Biol. 195 (7), 1071–1082.

Hershey, A.D., Chase, M., 1952. Independent functions of viral protein and nucleic acid in growth of bacteriophage. J. Gen. Physiol. 26, 36–56.

Kasamatsu, H., Nakanishi, A., 1998. How do animal DNA viruses get to the nucleus? Annu. Rev. Microbiol. 52, 627–686.

Lopez, S., Arias, C.F., 2004. Multistep entry of rotavirus into cells: a Versaillesque dance. Trends Microbiol. 12, 271–278.

Moore, J.P., et al., 2004. The CCR5 and CXCR4 coreceptors—central to understanding the transmission and pathogenesis of human immunodeficiency virus type 1 infection. AIDS Res. Hum. Retroviruses 20, 111–126.

Rossmann, M.G., et al., 2002. Picornavirus–receptor interactions. Trends Microbiol. 10, 324–331.

Sattentau, Q., 2008. Avoiding the void: cell-to-cell spread of human viruses. Nat. Rev. Microbiol. 6 (11), 815–826.

Welsch, S., Müller, B., Kräusslich, H.G., 2007. More than one door—Budding of enveloped viruses through cellular membranes. FEBS Lett. 581 (11), 2089–2097.

Wilen, C.B., Tilton, J.C., Doms, R.W., 2012. HIV: cell binding and entry. Cold Spring Harb. Perspect. Med. 2 (8), a006866.

Yamauchi, Y., Helenius, A., 2013. Virus entry at a glance. J. Cell Sci. 126 (6), 1289–1295.

CHAPTER 5

Expression

Intended Learning Outcomes

On completing this chapter you should be able to:

- Discuss the mechanisms by which cells express the information stored in genes.
- Describe the genome coding strategies of different virus groups.
- Explain how viruses control gene expression via transcription and by using posttranscriptional mechanisms.

EXPRESSION OF GENETIC INFORMATION

As described in Chapter 1, no virus yet discovered has the genetic information that encodes the tools necessary for the generation of metabolic energy, or for protein synthesis (ribosomes). So all viruses are dependent on their host cells for these functions, but the way in which viruses persuade their hosts to express their genetic information for them varies considerably. Patterns of virus replication are determined by tight controls on virus gene expression. There are fundamental differences in the control mechanisms of these processes in prokaryotic and eukaryotic cells, and these differences inevitably affect the viruses that utilize them as hosts. In addition, the relative simplicity and compact size of most virus genomes (compared with those of cells) creates further limits. Cells have evolved varied and complex mechanisms for controlling gene expression by using their extensive genetic capacity. Viruses have had to achieve highly specific quantitative, temporal, and spatial control of expression with much more limited genetic resources. Viruses have counteracted their genetic limitations by the evolving of a range of solutions to these problems. These mechanisms include:

- Powerful positive and negative signals that promote or repress gene expression.
- Highly compressed genomes in which overlapping reading frames are common.

CONTENTS

135

Principles of Molecular Virology. DOI: http://dx.doi.org/10.1016/B978-0-12-801946-7.00005-5

- Control signals that are frequently nested within other genes.
- Strategies which allow multiple polypeptides to be created from a single messenger RNA.

Gene expression involves regulatory loops mediated by signals that act either in *cis* (affecting the activity of neighboring genetic regions) or in *trans* (giving rise to diffusible products that act on regulatory sites any-where in the genome). For example, transcription **promoters** are *cis*-acting sequences that are located adjacent to the genes whose transcription they control, while proteins such as "transcription factors" which bind to specific sequences present on any stretch of nucleic acid present in the cell are examples of *trans*-acting factors. The relative simplicity of virus **genomes** and the elegance of their control mechanisms are models that form the basis of our current understanding of genetic regulation. This chapter assumes that you are familiar with the mechanisms involved in cellular control of gene expression. However, before we get into the details of virus gene expression, we will start with a brief reminder of some important aspects.

BOX 5.1 IT'S ALL ABOUT THE GENE

Even before Richard Dawkins wrote "The Selfish Gene" in the 1970s, the molecular biol-ogy revolution in the 1960s had ensured that the gene became the biological object that our thinking revolved around. In Dawkins view, genes are only concerned with their own sur-vival. When we look at how malleable virus genomes are, and how easily genes flow from host to virus and from one virus to another, you can understand why this view has become common. In many ways, this chapter on gene expression is at the very heart of this book. Understanding the mechanisms of transmis-sion and expression which act on genes is central to understanding modern biology.

CONTROL OF PROKARYOTE GENE EXPRESSION

Bacterial cells are second only to viruses in the specificity and economy of their genetic control mechanisms. In bacteria, genetic control operates both at the level of transcription and at subsequent (posttranscriptional) stages of gene expression.

The initiation of transcription is regulated primarily in a negative way by the synthesis of *trans*-acting repressor proteins, which bind to operator sequences upstream of protein-coding sequences. Collections of meta-bolically related genes are grouped together and coordinately controlled as "operons." Transcription of these operons typically produces a polycistronic **mRNA** that encodes several different proteins. During subsequent stages of expression, transcription is also regulated by a number of mechanisms that act, in Mark Ptashne's famous phrase, as "genetic switches," turning on or off the transcription of different genes. Such mechanisms include antitermination, which is controlled by *trans*-acting factors that promote the synthesis of longer transcripts encoding additional genetic information, and by various modifications of RNA polymerase. Bacterial σ (sigma) factors are apoproteins that affect the specificity of the RNA polymerase holoenzyme (active form) for different promoters. Several bacteriophages (e.g., phage SP01 of *Bacillus subtilis*) encode proteins that function as alternative σ factors, sequestering RNA polymerase and altering the rate at which phage genes are transcribed. Phage T4 of *Escherichia coli* encodes an enzyme that carries out a covalent modification (adenosine diphosphate [ADP]-ribosylation) of the host-cell RNA polymerase. This is believed to eliminate the requirement of the polymerase holoenzyme for σ factor and to achieve an effect similar to the production of modified σ factors by other bacteriophages.

At a posttranscriptional level, gene expression in bacteria is also regulated by control of translation. The best-known virus examples of this phenomenon come from the study of bacteriophages of the family *Leviviridae*, such as R17, MS2, and Qβ. In these phages, the secondary structure of the single-stranded RNA phage genome not only regulates the quantities of different phage proteins that are translated but also operates temporal (timed) control of a switch in the ratios between the different proteins produced in infected cells.

CONTROL OF EXPRESSION IN BACTERIOPHAGE λ

The genome of phage λ has been studied in great detail and illustrates several of the mechanisms described above, including the action of repressor proteins in regulating lysogeny versus lytic replication and antitermination of transcription by phage-encoded *trans*-acting factors. Such has been the impact of these discoveries that no discussion of the control of virus gene expression is complete without detailed examination of this phage.

Phage λ was discovered by Esther Lederberg in 1949. Experiments at the Pasteur Institute by André Lwoff in 1950 showed that some strains of *Bacillus megaterium*, when irradiated with ultraviolet light, stopped growing and subsequently lysed, releasing a crop of bacteriophage particles. Together with Francois Jacob and Jacques Monod, Lwoff subsequently showed that the cells of some bacterial strains carried a bacteriophage in a dormant form, known as a prophage, and that the phage could be made to alternate between the lysogenic (nonproductive) and lytic (productive) growth cycles.

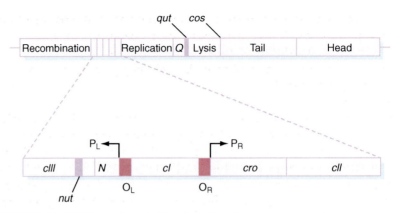

FIGURE 5.1 Simplified genetic map of bacteriophage λ.
The top part of this figure shows the main genetic regions of the phage genome and the bottom part is an expanded view of the main control elements described in the text.

After many years of study, our understanding of λ has been refined into a picture that represents one of the best understood and most elegant genetic control systems yet to be investigated. A simplified genetic map of λ is shown in Figure 5.1.

For regulation of the growth cycle of the phage, the structural genes encoding the head and tail components of the virus **capsid** can be ignored. The components involved in genetic control are as follows:

1. P_L is the promoter responsible for transcription of the left-hand side of the λ genome, including *N* and *cIII*.
2. O_L is a short noncoding region (NCR) of the phage genome (approximately 50 bp) which lies between the *cI* and *N* genes next to P_L.
3. P_R is the promoter responsible for transcription of the right-hand side of the λ genome, including *cro*, *cII*, and the genes encoding the structural proteins.
4. O_R is a short NCR of the phage genome (approximately 50 bp) which lies between the *cI* and *cro* genes next to P_R.
5. *cI* is transcribed from its own promoter and encodes a repressor protein of 236 amino acids which binds to O_R, preventing transcription of *cro* but allowing transcription of *cI*, and to O_L, preventing transcription of *N* and the other genes at the left-hand end of the genome.
6. *cII* and *cIII* encode activator proteins that bind to the genome, enhancing the transcription of the *cI* gene.
7. *cro* encodes a 66-amino acid protein that binds to O_R, blocking binding of the repressor to this site.

8. *N* encodes an antiterminator protein that acts as an alternative ρ (rho) factor for host-cell RNA polymerase, modifying its activity and permitting extensive transcription from P_L and P_R.

9. *Q* is an antiterminator similar to *N*, but it only permits extended transcription from P_R.

In a newly infected cell, *N* and *cro* are transcribed from P_L and P_R, respectively (Figure 5.2). The N protein allows RNA polymerase to transcribe a number of phage genes, including those responsible for DNA **recombination** and integration of the prophage, as well as cII and cIII. The N protein acts as a positive transcription regulator. In the absence of the N protein, the RNA polymerase holoenzyme stops at certain sequences located at the end of the *N* and *Q* genes, known as the *nut* and *qut* sites, respectively. However, RNA polymerase−N protein complexes are able to overcome this restriction and permit full transcription from P_L and P_R. The RNA polymerase−Q protein complex results in extended transcription from P_R only. As levels of the cII and cIII proteins in the cell build up, transcription of the *cI* repressor gene from its own promoter is turned on.

At this point, the critical event that determines the outcome of the infection occurs. The cII protein is constantly degraded by host-cell proteases. If levels of cII remain below a critical level, transcription from P_R and P_L continues, and the phage undergoes a productive replication cycle that culminates in lysis of the cell and the **release** of phage particles. This is the sequence of events that occurs in the vast majority of infected cells. However, in a few rare instances, the concentration of cII protein builds up, transcription of *cI* is enhanced, and intracellular levels of the cI repressor protein rise. The repressor binds to O_R and O_L which prevents transcription of all phage genes (particularly *cro*; see below) except itself. The level of cI protein is maintained automatically by a negative feedback mechanism, as at high concentrations the repressor also binds to the left-hand end of O_R and prevents transcription of *cI* (Figure 5.3). This autoregulation of cI synthesis keeps the cell in a stable state of **lysogeny**.

If this is the case, how do such cells ever leave this state and enter a productive, **lytic** replication cycle? Physiological stress and particularly ultraviolet irradiation of cells result in the induction of a host-cell protein, RecA. This protein, the normal function of which is to induce the expression of cellular genes that permit the cell to adapt to and survive in altered environmental conditions, cleaves the cI repressor protein. In itself, this would not be sufficient to prevent the cell from reentering the lysogenic state; however, when repressor protein is not bound to O_R, *cro* is transcribed from P_R. Cro also binds to O_R but, unlike cI, which preferentially binds the right-hand end of O_R, the Cro protein binds preferentially to the left-hand end of O_R, preventing the transcription of *cI* and enhancing its own transcription in a positive-feedback loop. The phage is then locked into a lytic cycle and cannot return to the lysogenic state.

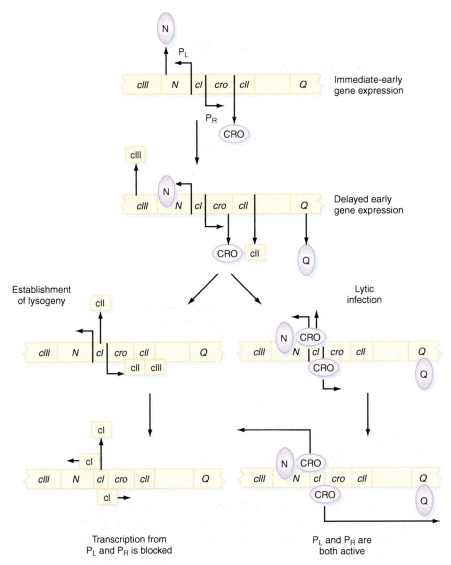

FIGURE 5.2 Control of expression of the bacteriophage λ genome.
See text for a detailed description of the events that occur in a newly infected cell and during lytic infection or lysogeny.

The molecular details of λ gene have contributed greatly to our understanding of genetic regulation in **prokaryotic** and **eukaryotic** cells. Determination of the structures of the proteins involved in the above scheme has allowed us to identify the fundamental principles behind the observation that many proteins from unrelated organisms can recognize and bind to specific sequences in DNA molecules. The concepts of proteins with independent DNA-binding and dimerization domains, protein cooperativity in DNA

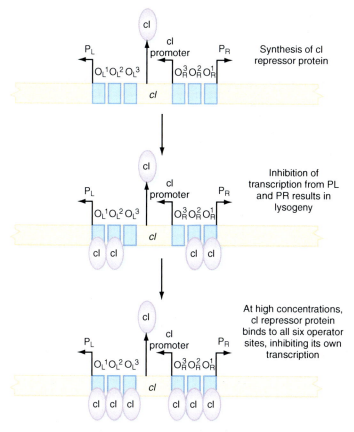

FIGURE 5.3 Control of lysogeny in bacteriophage λ.
See text for a detailed description of the events that occur in the establishment and maintenance of lysogeny.

binding, and DNA looping allowing proteins bound at distant sites to interact with one another have all risen from the study of λ. The references given at the end of this chapter explain more fully the nuances of gene expression in this complex bacteriophage.

BOX 5.2 BACTERIOPHAGES ARE LIKE, SO LAST CENTURY

No they're not. Apart from the contribution of bacteriophages to understanding viruses as a whole—and there's no better example of that than λ—some of the most exciting work in virology over the last decade is about phages. When we finally raised our sights from the glassware in our laboratories and went out hunting for viruses in the natural environment, we were staggered at what we found. Phages in particular are everywhere, and in staggering quantities and variety. It has been estimated that every second on Earth, 1×10^{25} bacteriophage infections occur. That means that phages control the turnover of such large quantities of organic material that this has a major impact on nutrient cycling and the global climate. Last century? Wrong.

CONTROL OF EUKARYOTE GENE EXPRESSION

Control of gene expression in eukaryotic cells is much more complex than in prokaryotic cells and involves a multilayered approach in which diverse control mechanisms exert their effects at different levels. The first level of control occurs prior to transcription and depends on the local configuration of the DNA. DNA in eukaryotic cells has an elaborate structure, forming complicated and dynamic but far from random complexes with numerous proteins to form chromatin. Although the contents of eukaryotic cell nuclei appear amorphous in electron micrographs (at least in interphase), they are actually highly ordered. Chromatin interacts with the structural backbone of the nucleus—the nuclear matrix—and these interactions are thought to be important in controlling gene expression. Locally, nucleosome configuration and DNA conformation, particularly the formation of left-handed helical "Z-DNA," are also important. DNAse I digestion of chromatin does not give an even, uniform digestion pattern but reveals a pattern of DNAse hyper-sensitive sites believed to indicate differences in the function of various regions of the chromatin. It is likely, for example, that retroviruses are more likely to integrate into the host-cell genome at these sites than elsewhere. Transcriptionally active DNA is also hypomethylated—that is, there is a rela-tive scarcity of nucleotides modified by the covalent attachment of methyl groups in these regions compared with the frequency of methylation in transcriptionally quiescent regions of the genome. The methylation of inte-grated retrovirus genomes has been shown to suppress the transcription of the provirus genome.

The second level of control rests in the process of transcription itself, which again is much more complex than in prokaryotes. There are three forms of RNA polymerase in eukaryotic cells that can be distinguished by their relative sensitivity to the drug α-amanitin and which are involved in the expression of different classes of genes (Table 5.1). The rate at which tran-scription is initiated is a key control point in eukaryotic gene expression. Initiation is influenced dramatically by sequences upstream of the transcrip-tion start site which function by acting as recognition sites for families of highly specific DNA-binding proteins, known as "transcription factors."

Table 5.1 Forms of RNA Polymerase in Eukaryotic Cells

RNA Polymerase	Sensitivity to α-Amanitin	Cellular Genes Transcribed	Virus Genes Transcribed
I	Unaffected	Ribosomal RNAs	—
II	Highly sensitive	Most single-copy genes	Most DNA virus genomes
III	Moderately sensitive	5S rRNA, tRNAs	Adenovirus VA RNAs

Immediately upstream of the transcription start site is a relatively short region known as the promoter. It is at this site that transcription complexes, consisting of RNA polymerase plus accessory proteins, bind to the DNA and transcription begins. However, sequences further upstream from the promoter also influence the efficiency with which transcription complexes form. The rate of initiation depends on the combination of transcription factors bound to these transcription enhancers. The properties of these enhancer sequences are remarkable in that they can be inverted and/or moved around relative to the position of the transcription start site without losing their activity and can exert their influence even from a distance of several kilobases away. This emphasizes the flexibility of DNA, which allows proteins bound at distant sites to interact with one another, as also shown by the protein–protein interactions seen in regulation of phage λ gene expression (above). Transcription of eukaryotic genes results in the production of monocistronic mRNAs, each of which is transcribed from its own individual promoter.

At the next stage, gene expression is influenced by the structure of the mRNA produced. The stability of eukaryotic mRNAs varies considerably, some having comparatively long half-lives in the cell (e.g., many hours). The half-lives of others, typically those that encode regulatory proteins, may be very short (e.g., a few minutes). The stability of eukaryotic mRNAs depends on the speed with which they are degraded. This is determined by such factors as its terminal sequences, which consist of a methylated cap structure at the 5′ end and polyadenylic acid at the 3′ end, as well as on the overall secondary structure of the message. However, gene expression is also regulated by differential splicing of heterogeneous (heavy) nuclear RNA (hnRNA) precursors in the nucleus, which can alter the genetic meaning of different mRNAs transcribed from the same gene. In eukaryotic cells, control is also exercised during export of RNA from the nucleus to the cytoplasm.

Finally, the process of translation offers further opportunities for control of expression. The efficiency with which different mRNAs are translated varies greatly. These differences result largely from the efficiency with which ribosomes bind to different mRNAs and recognize AUG translation initiation codons in different sequence contexts, as well as the speed at which different sequences are converted into protein. Certain sequences act as translation enhancers, performing a function analogous to that of transcription enhancers.

MicroRNAs (miRNAs) are small (approximately 22 nt) RNAs that play important roles in the regulation of gene expression, typically silencing gene expression by directing repressive protein complexes to the 3′ untranslated region of specific target messenger mRNA transcripts. First discovered to play important roles in posttranscriptional gene regulation in eukaryotic cells, many virus-encoded miRNAs are now known from diverse virus families, most with DNA genomes but also some RNA viruses, including human immunodeficiency virus (HIV) and Ebola virus. Some of the best characterized virus miRNA functions include prolonging the longevity of infected cells (several different herpesviruses), evading the immune response (in polyomaviruses), and regulating the switch to lytic infection (herpesviruses). This enables gene expression to continue in infected cells in circumstances where in the absence of the miRNAs it would be shut down.

The point of this extensive list of eukaryotic gene expression mechanisms is that they are all utilized by viruses to control gene expression. Examples of each type are given in the sections below. If this seems remarkable, remember that the control of gene expression in eukaryotic cells was unraveled largely by utilizing viruses as model systems, therefore finding examples of these mechanisms in viruses is really only a self-fulfilling prophecy.

In recent years, the importance of epigenetics in shaping virus gene expression has been recognized. Epigenetics refers to control of gene expression independent of changes in the DNA sequence of the gene affected. Some of the processes involved include DNA methylation, histone modifications, chromatin-remodeling proteins, and DNA silencing. In viruses, examples of epigenetic mechanisms controlling gene expression include influenza virus NS1 protein mimicking histone H3, chromatin assembly and histone and DNA modifications in the control of latent infections of herpesviruses, and the importance of epigenetics in controlling virus-associated (VA) tumor formation.

As discussed in Chapter 1, advances in biotechnology such as next generation sequencing methods have had a major impact on the understanding of virus replication. In understanding gene expression, rapid direct RNA sequencing methods ("RNAseq") has been particularly important. Microarray technology evolved from Southern blotting. Tens of thousands of DNA probes, usually chemically synthesized oligonucleotides, can be attached to a glass slide and the genes they represent can all be analyzed in a single experiment. To achieve the very high density of DNA probes in the array, spotting or printing of the slides is carried out by specialized robots. DNA microarrays are most commonly used to detect mRNAs and this is referred to as expression analysis or expression profiling. The cyanine dyes Cy3 and Cy5 which have excitation wavelengths of 635 nm and 532 nm respectively are most frequently used to label the RNA samples. Two-color labeling allows two samples, for example infected and uninfected cells, to be hybridized to the same array and their gene expression profiles compared via the difference in the fluorescence of the two samples. This complex mixture of sequences is then allowed to hybridize to the DNA capture probes on the array. Unbound (noncomplementary) sequences are washed away and the fluorescence of the individual

DNA spots on the array is measured by laser excitation. Statistical processing of the fluorescence data is usually necessary to eliminate artifacts and false results from the data obtained. Because of the power of this technique, microarrays have become the preeminent technology for the investigation of functional genomics—the area of modern biology which focuses on dynamic aspects such as gene transcription, translation, and protein–protein interactions.

GENOME CODING STRATEGIES

BOX 5.3 SO MANY VIRUSES, HOW AM I GOING TO REMEMBER THEM ALL?

Good question. You don't need to remember all the details about every virus—even people who write virology textbooks can't do that. What you do need to do is to have a framework which allows you to think "Yes, I've seen something like this before, so I can guess what's likely to happen." And that's where the seven classes of virus genomes described in the previous chapter come in. Add on to that an understanding of how gene expression works for each type and you're pretty much there. There's one small catch. Even for viruses with very similar genome structures, there are often surprising differences in mechanisms of gene expression. Biology is all about variation. If you wanted everything to be predictable, you should have signed up for the physics class!

In Chapter 4, genome structure was one element of a classification scheme used to divide viruses into seven groups. The other part of this scheme is the way in which the genetic information of each class of virus genomes is expressed. The replication and expression of virus genomes are inseparably linked, and this is particularly true in the case of RNA viruses. Here, the seven classes of virus genome described in Chapter 4 and Appendix 1 are reviewed again, this time examining the way in which the genetic information of each class is expressed.

Class I: Double-Stranded DNA

Chapter 4 said that this class of virus genomes can be subdivided into two further groups: those in which genome replication is exclusively nuclear (e.g., *Adenoviridae*, *Polyomaviridae*, *Herpesviridae*) and those in which replication occurs in the cytoplasm (*Poxviridae*). In one sense, all of these viruses can be considered to be similar—because their genomes all resemble double-stranded cellular DNA, and they are essentially transcribed by the same mechanisms as cellular genes. However, there are profound differences between them relating to the degree to which each family is reliant on the host-cell machinery.

Polyomaviruses and Papillomaviruses

Polyomaviruses are heavily dependent on cellular machinery for both replication and gene expression. Polyomaviruses encode *trans*-acting factors (T-antigens) which stimulate transcription (and genome replication). Papillomaviruses in particular are dependent on the cell for replication, which only occurs in terminally differentiated keratinocytes and not in other cell types, although they do encode several *trans*-regulatory proteins (Chapter 7).

Adenoviruses

Adenoviruses are also heavily dependent on the cellular apparatus for transcription, but they possess various mechanisms that specifically regulate virus gene expression. These include *trans*-acting transcriptional activators such as the E1A protein, and posttranscriptional regulation of expression, which is achieved by alternative splicing of mRNAs and the virus-encoded VA RNAs (Chapter 7). Adenovirus infection of cells is divided into two stages, early and late, the latter phase commencing at the time when genome replication occurs; however, in adenoviruses, these phases are less distinct than in herpesviruses (below).

Herpesviruses

These viruses are less reliant on cellular enzymes than the previous groups. They encode many enzymes involved in DNA metabolism (e.g., thymidine kinase) and a number of *trans*-acting factors that regulate the temporal expression of virus genes, controlling the phases of infection. Transcription of the large, complex genome is sequentially regulated in a cascade fashion (Figure 5.4). At least 50 virus-encoded proteins are produced after transcription of the genome by host-cell RNA polymerase II. Three distinct classes of mRNAs are made:

- α: Immediate-early (IE) mRNAs encode *trans*-acting regulators of virus transcription.
- β: (Delayed) early mRNAs encode further nonstructural regulatory proteins and some minor structural proteins.
- γ: Late mRNAs encode the major structural proteins.

Gene expression in herpesviruses is tightly and coordinately regulated, as indicated by the following observations (see Figure 5.4). If translation is blocked shortly after infection (e.g., by treating cells with cycloheximide), the production of late mRNAs is blocked. Synthesis of the early gene product turns off the IE products and initiates genome replication. Some of the late structural proteins (γ1) are produced independently of genome replication; others (γ2) are only produced after replication. Both the IE and early proteins are required to initiate genome replication. A virus-encoded DNA-dependent DNA polymerase and a DNA-binding protein are involved in

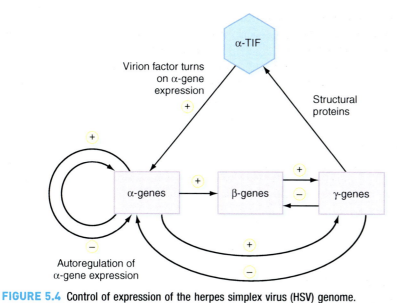

FIGURE 5.4 **Control of expression of the herpes simplex virus (HSV) genome.**
HSV particles contain a protein called α-gene transcription initiation factor (α-TIF) which turns on
α-gene expression in newly infected cells, beginning a cascade of closely regulated events which
control the expression of the entire complement of the 70 or so genes in the virus genome.

genome replication, together with a number of enzymes (e.g., thymidine
kinase) that alter cellular biochemistry. The production of all of these proteins is closely controlled.

Poxviruses

Genome replication and gene expression in poxviruses are almost independent of cellular mechanisms (except for the requirement for host-cell ribosomes). Poxvirus genomes encode numerous enzymes involved in DNA
metabolism, virus gene transcription, and posttranscriptional modification of
mRNAs. Many of these enzymes are packaged within the virus particle
(which contains >100 proteins), enabling transcription and replication of
the genome to occur in the cytoplasm (rather than in the nucleus, like all the
families described above) almost totally under the control of the virus. Gene
expression is carried out by virus enzymes associated with the core of the
particle and is divided into two rather indistinct phases:

- **Early genes:** These comprise about 50% of the poxvirus genome and
  are expressed before genome replication inside a partially uncoated
  core particle (Chapter 2), resulting in the production of 5′ capped,
  3′ polyadenylated but unspliced mRNAs.

■ **Late genes:** These are expressed after genome replication in the cytoplasm, but their expression is also dependent on virus-encoded rather than on cellular transcription proteins (which are located in the nucleus). Like herpesviruses, late gene promoters are dependent on prior DNA replication for activity.

The Giant Viruses: Mimivirus, Megavirus, Pandoravirus, Pithovirus

As might be expected from the trend seen in the above families, the giant dsDNA viruses replicate in the cytoplasm, forming dense vesicles which act as virus factories. Some of these virus genomes contain large numbers of non-coding repeated elements in a similar way to eukaryotic genomes. Of the 2,500 predicted proteins of *Pandoravirus*, most of them are unique and not found in other viruses or cellular genomes. Typically, the coding genes are expressed as monocistronic mRNAs with very short NCRs at each end, shorter than the ones found in eukaryotic transcripts, leaving little room for regulatory signals in the transcript.

Class II: Single-Stranded DNA

Both the autonomous and the helper virus-dependent parvoviruses are highly reliant on host-cell assistance for gene expression and genome replication. This is presumably because the very small size of their genomes does not permit them to encode the necessary biochemical apparatus. These viruses show an extreme form of parasitism, utilizing the normal functions present in the nucleus of their host cells for both expression and replication (Figure 5.5). The members of the replication-defective *Dependovirus* genus of the *Parvoviridae* are entirely dependent on adenovirus or herpesvirus superinfection for the provision of further helper functions essential for their replication beyond those present in normal cells. The adenovirus genes required as helpers are the early, transcriptional regulatory genes such as E1A rather than late structural genes, but it has been shown that treatment of cells with ultraviolet light, cycloheximide, or some carcinogens can replace the requirement for helper viruses. Therefore, the help required appears to be for a modification of the cellular environment (probably affecting transcription of the defective parvovirus genome) rather than for a specific virus protein.

The *Geminiviridae* also fall into this class of genome structures (Figure 3.15). The expression of their genomes is quite different from that of parvoviruses but nevertheless still relies heavily on host-cell functions. There are open reading frames in both orientations in the virus DNA, which means that both (+) and (−)sense strands are transcribed during infection. The mechanisms involved in control of gene expression have not been fully investigated, but at least some geminiviruses (subgroup I) may use splicing.

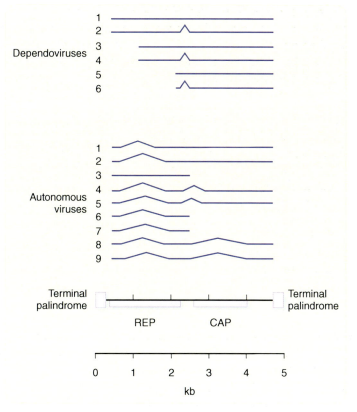

FIGURE 5.5 Transcription of parvovirus genomes.
Transcription of parvovirus genomes is heavily dependent on host-cell factors and results in the synthesis of a series of spliced, subgenomic mRNAs that encode two proteins: Rep, which is involved in genome replication, and Cap, the capsid protein (see text).

Class III: Double-Stranded RNA

All viruses with RNA genomes differ fundamentally from their host cells, which of course possess double-stranded DNA genomes. Therefore, although each virus must be biochemically "compatible" with its host cell, there are fundamental differences in the mechanisms of virus gene expression from those of the host cell. Reoviruses have multipartite genomes (see Chapter 3) and replicate in the cytoplasm of the host cell. Characteristically for viruses with segmented RNA genomes, a separate **monocistronic** mRNA is produced from each segment (Figure 5.6). Early in infection, transcription of the **dsRNA** genome segments by virus-specific **transcriptase** activity occurs inside partially uncoated subvirus particles. At least five enzymatic activities are present in reovirus particles to carry out this process, although these are not necessarily all separate peptides (Table 5.2; Figure 2.12). This primary transcription results in

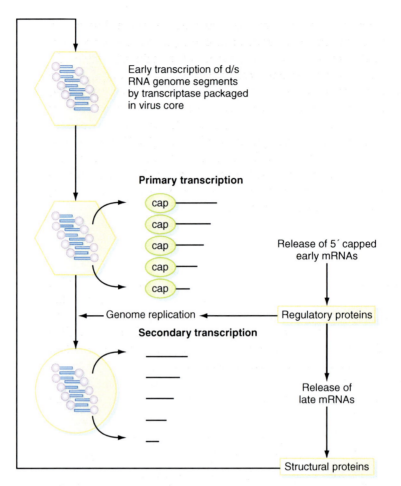

Early transcription of d/s
RNA genome segments
by transcriptase packaged
in virus core

Primary transcription

cap ————
cap ————
cap ————
cap ————
cap ————

Release of 5′ capped
early mRNAs

← Genome replication ←   Regulatory proteins

Secondary transcription

Release of
late mRNAs

Structural proteins

FIGURE 5.6 Expression of reovirus genomes.
Expression of reovirus genomes is initiated by a transcriptase enzyme packaged inside every virus particle. Subsequent events occur in a tightly regulated pattern, the expression of late mRNAs encoding the structural proteins being dependent on prior genome replication.

Table 5.2 Enzymes in Reovirus Particles

Activity	Virus Protein	Encoded by Genome Segment
dsRNA-dependent RNA polymerase (Pol)	λ3	L3
RNA triphosphatase	μ2	M1
Guanyltransferase (Cap)	λ2	L2
Methyltransferase	λ2	L2
Helicase (Hel)	λ1	L1

capped transcripts that are not polyadenylated and which leave the virus core to be translated in the cytoplasm. The various genome segments are transcribed/translated at different frequencies, which is perhaps the main advantage of a segmented genome. RNA is transcribed conservatively; that is, only (−)sense strands are used, resulting in synthesis of (+)sense mRNAs, which are capped inside the core (all this occurs without *de novo* protein synthesis). Secondary transcription occurs later in infection inside new particles produced in infected cells and results in uncapped, nonpolyadenylated transcripts. The genome is replicated in a conservative fashion (cf. semiconservative DNA replication). Excess (+)sense strands are produced which serve as late mRNAs and as templates for (−)sense strand synthesis (i.e., each strand leads to many (+) strands, not one-for-one as in semiconservative replication).

Class IV: Single-Stranded (+)Sense RNA

This type of genome occurs in many animal viruses and plant viruses (Appendix 2). In terms of both the number of different families and the number of individual viruses, this is the largest single class of virus genomes. Essentially, these virus genomes act as messenger RNAs and are themselves translated immediately after infection of the host cell (Chapter 3). Not surprisingly with so many representatives, this class of genomes displays a very diverse range of strategies for controlling gene expression and genome replication. However, in very broad terms, the viruses in this class can be subdivided into two groups:

1. Production of a **polyprotein** encompassing the whole of the virus genetic information, which is subsequently cleaved by proteases to produce precursor and mature polypeptides. These cleavages can be a subtle way of regulating the expression of genetic information. Alternative cleavages result in the production of various proteins with distinct properties from a single precursor (e.g., in picornaviruses and potyviruses) (Figure 5.7). Certain plant viruses with multipartite genomes utilize a very similar strategy for controlling gene expression, although a separate polyprotein is produced from each of the genome segments. The best-studied example of this is the comoviruses, whose genome organization is very similar to that of the picornaviruses and may represent another member of this "superfamily" (Figure 5.7).

2. Production of subgenomic mRNAs, resulting from two or more rounds of translation of the genome. This strategy is used to achieve temporal separation of what are essentially "early" and "late" phases of replication, in which nonstructural proteins, including a virus **replicase**, are produced during the "early" phase followed by structural proteins in the "late" phase (Figure 5.8). The proteins produced in

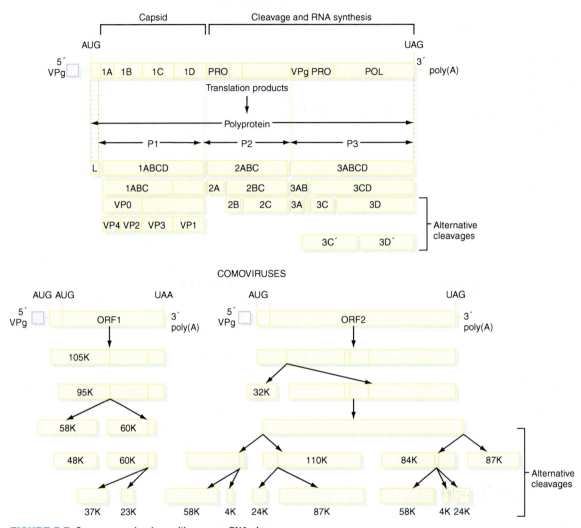

FIGURE 5.7 Gene expression in positive-sense RNA viruses.
Many positive-sense RNA virus genomes are frequently translated to form a long polyprotein, which is subsequently cleaved by a highly specific virus-encoded protease to form the mature polypeptides. Picornaviruses and comoviruses are examples of this mechanism of gene expression.

each of these phases may result from proteolytic processing of a polyprotein precursor, although this encompasses only part of the virus genome rather than the entire genome, as above. Proteolytic processing offers further opportunities for regulation of the ratio of different polypeptides produced in each phase of replication

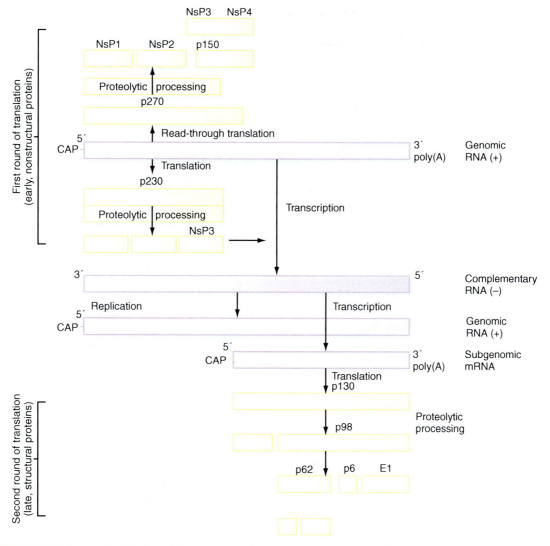

FIGURE 5.8 Subgenomic RNAs in positive-sense RNA viruses.
Some positive-sense RNA virus genomes (e.g., togaviruses) are expressed by two separate rounds of translation, involving the production of a subgenomic mRNA at a later stage of replication.

(e.g., in togaviruses and tymoviruses). In addition to proteolysis, some viruses employ another strategy to produce alternative polypeptides from a subgenomic mRNA, either by read-through of a "leaky" translation stop codon (e.g., tobamoviruses such as TMV; see Figure 3.12), or by deliberate ribosomal frameshifting at a particular site (see "Posttranscriptional Control of Expression").

All the viruses in this class have evolved mechanisms that allow them to regulate their gene expression in terms of both the ratios of different virus-
encoded proteins and the stage of the replication cycle when they are
produced. Compared with the two classes of DNA virus genomes described
above, these mechanisms operate largely independently of those of the host
cell. The power and flexibility of these strategies are reflected very clearly in
the overall success of the viruses in this class, as determined by the number of
different representatives known and the number of different hosts they infect.

Class V: Single-Stranded (−)Sense RNA

As discussed in Chapter 3, the genomes of these viruses may be either
segmented or nonsegmented. The first step in the replication of segmented
orthomyxovirus genomes is transcription of the (−)sense vRNA by the virion-
associated RNA-dependent RNA polymerase to produce (predominantly)
monocistronic mRNAs, which also serve as the template for subsequent
genome replication (Figure 5.9). As with all (−)sense RNA viruses, packaging
of this virus-specific transcriptase/replicase within the virus nucleocapsid is
essential because no host cell contains any enzyme capable of decoding and
copying the RNA genome. Atypically for RNA viruses, influenza and its close
relatives replicate in the nucleus of the host cell rather than the cytoplasm.
This allows these viruses to use the splicing apparatus present in this compartment to express subgenomic messages from some, although not all, of its
genome segments.

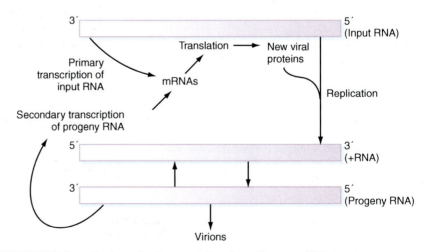

FIGURE 5.9 General scheme for the expression of negative-sense RNA virus genomes.
All negative-sense RNA viruses face the problem that the information stored in their genome cannot be
interpreted directly by the host cell. They must include mechanisms for converting the genome into
mRNAs within the virus particle.

In the other families that have nonsegmented genomes, monocistronic mRNAs are also produced. However, these messages must be produced from a single, long (−)sense RNA molecule. Exactly how this is achieved is not clear. It is possible that a single, genome-length transcript is cleaved after transcription to form the separate mRNAs, but it is more likely that these are produced individually by a stop-and-start mechanism of transcription regulated by the conserved intergenic sequences present between each of the virus genes (Chapter 3). Splicing mechanisms cannot be used because these viruses replicate in the cytoplasm.

On the surface, such a scheme of gene expression might appear to offer few opportunities for regulation of the relative amounts of different virus proteins. If this were true, it would be a major disadvantage, as all viruses require far more copies of the structural proteins (e.g., nucleocapsid protein) than of the nonstructural proteins (e.g., polymerase) for each virion produced. In practice, the ratio of different proteins is regulated both during transcription and afterwards. In paramyxoviruses, for example, there is a clear polarity of transcription from the 3′ end of the virus genome to the 5′ end which results in the synthesis of far more mRNAs for the structural proteins encoded in the 3′ end of the genome than for the nonstructural proteins located at the 5′ end (Figure 5.10). Similarly, the advantage of producing monocistronic mRNAs is that the translational efficiency of each message can be varied with respect to the others (see "Posttranscriptional Control of Expression").

Viruses with ambisense genome organization (where genetic information is encoded in both the positive (i.e., virus-sense) and negative (i.e., complementary) orientations on the same strand of RNA; see Chapter 4) must express their genes in two rounds of expression so that both are turned into decodable mRNA at some point (Figure 5.11).

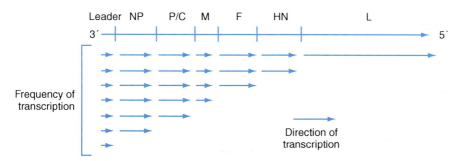

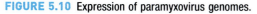

FIGURE 5.10 Expression of paramyxovirus genomes.
Paramyxovirus genomes exhibit "transcriptional polarity." Transcripts of genes at the 3′ end of the virus genome are more abundant than those of genes at the 5′ end of the genome, permitting regulation of the relative amounts of structural (3′ genes) and nonstructural (5′ genes) proteins produced.

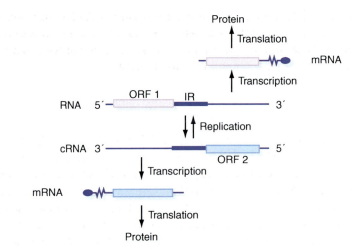

FIGURE 5.11 Expression of ambisense virus genomes.
Ambisense virus genomes (which have both positive and negative-sense information in the same strand of RNA) require two rounds of gene expression so that information encoded in both strands of the genome is turned into decodable mRNA at some point in the replication cycle.

Class VI: Single-Stranded (+)Sense RNA with DNA Intermediate

The retroviruses are the ultimate case of reliance on the host-cell transcription machinery. The RNA genome forms a template for reverse transcription to DNA—these are the only (+)sense RNA viruses whose genome does not serve as mRNA on entering the host cell (Chapter 3). Once integrated into the host-cell genome, the DNA provirus is under the control of the host cell and is transcribed exactly as are other cellular genes. Some retroviruses, however, have evolved a number of transcriptional and posttranscriptional mechanisms that allow them to control the expression of their genetic information, and these are discussed in detail later in this chapter.

Class VII: Double-Stranded DNA with RNA Intermediate

Expression of the genomes of these viruses is complex and relatively poorly understood. The hepadnaviruses contain a number of overlapping reading frames clearly designed to squeeze as much coding information as possible into a compact genome. The Hepatitis B virus X gene encodes a transcriptional *trans*-activator believed to be analogous to the human T-cell leukemia virus (HTLV) tax protein (see later). At least two mRNAs are produced from independent promoters, each of which encodes several proteins and the larger of which is also the template for reverse transcription during the formation of the virus particle (Chapter 3). Expression of caulimovirus genomes is similarly complex, although there are similarities with hepadnaviruses in that two major transcripts are produced, 35S and 19S. Each of these encodes several polypeptides, and the 35S transcript is the template for reverse transcription during the formation of the virus genome.

TRANSCRIPTIONAL CONTROL OF EXPRESSION

Having looked at general strategies used by different groups of viruses to regulate gene expression, the rest of this chapter concentrates on more detailed explanations of specific examples from some of the viruses mentioned earlier, beginning with control of transcription in SV40, a member of the *Polyomaviridae*. Few other genomes, virus or cellular, have been studied in as much detail as SV40, which has been a model for the study of eukaryotic transcription mechanisms (particularly DNA replication; see Chapter 6) for many years. In this sense, SV40 provides a eukaryotic parallel with the bacteriophage λ genome. *In vitro* systems exist for both transcription and replication of the SV40 genome, and it is believed that all the virus and cellular DNA-binding proteins involved in both of these processes are known. The SV40 genome encodes two T-antigens ("tumor antigens") known as large T-antigen and small T-antigen after the sizes of the proteins (Figure 5.12). Replication of the double-stranded DNA genome of SV40 occurs in the nucleus of the host cell. Transcription of the genome is carried out by host-cell RNA polymerase II, and large T-antigen plays a vital role in regulating transcription of the virus genome. Small T-antigen is not essential for virus replication but does allow virus DNA to accumulate in the nucleus. Both proteins contain nuclear localization signals which result in their accumulation in the nucleus, where they migrate after being synthesized in the cytoplasm.

Soon after infection of permissive cells, early mRNAs are expressed from the early promoter, which contains a strong transcription enhancer element (the 72-bp sequence repeats), allowing it to be active in newly infected cells (Figure 5.13). The early proteins made are the two T-antigens. As the concentration of large T-antigen builds up in the nucleus, transcription of the early genes is repressed by direct binding of the protein to the origin region of the virus genome, preventing transcription from the early promoter and causing the switch to the late phase of infection. Large T-antigen is also required for replication of the genome, and this is discussed further in Chapter 7. After DNA replication has occurred, transcription of the late genes occurs from the late promoter and results in the synthesis of the structural proteins VP1, VP2, and VP3. This process illustrates two classic features of control of virus gene expression. First, the definition of the "early" and "late" phases of replication, when different sets of genes tend to be expressed, is before/after genome replication. Second, there is usually a crucial protein, in this case T-antigen,

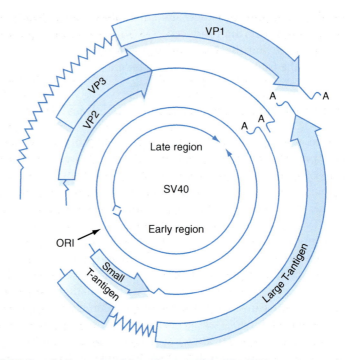

FIGURE 5.12 Organization and protein-coding potential of the SV40 genome.
The highly compact SV40 genome includes genetic information encoded in overlapping reading frames on both strands of the DNA genome and mRNA splicing to produce alternative polypeptides from one open reading frame.

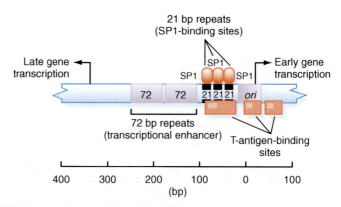

FIGURE 5.13 Control of transcription of the SV40 genome.
Multiple virus-encoded (T-antigen) and cellular proteins bind to the *ori* region of the SV40 genome to control gene expression during different phases of replication (see text for details).

whose function is comparable to that of a "switch"—you should compare the pattern of replication of SV40 with the description of bacteriophage λ gene expression control given earlier in this chapter.

Another area where control of virus transcription has received much attention is in the human retroviruses, HTLV, and HIV. Integrated DNA proviruses are formed by reverse transcription of the RNA retrovirus genome, as described in Chapter 3. The presence of numerous binding sites for cellular transcription factors in the long terminal repeats (LTRs) of these viruses have been analyzed by "DNAse I footprinting" and "gel-shift" assays (Figure 5.14). Together, the "distal" elements (such as NF-κB and SP1-binding sites) and "proximal" elements (such as the TATA box) make up a transcription promoter in the U3 region of the LTR (Chapter 3). However, the basal activity of this promoter on its own is relatively weak, and results in only limited transcription of the provirus genome by RNA polymerase II. Both HTLV and HIV encode proteins that are *trans*-acting positive regulators of transcription: the Tax protein of HTLV and the HIV Tat protein (Figure 5.15). These proteins act to increase transcription from the virus LTR by a factor of at least 50–100 times that of the basal rate from the "unaided" promoter. Unlike T-antigen and the early promoter of SV40, neither the Tax nor the Tat proteins (which have no structural similarity to one another) bind directly to its respective LTR. Instead, these proteins function indirectly by interacting with cellular transcription factors which in turn bind to the promoter region of the virus LTR. So the HTLV Tax and HIV Tat proteins are positive regulators of the basal promoter in the provirus LTR and are under the control of the virus, as synthesis of these proteins is dependent on the promoters which

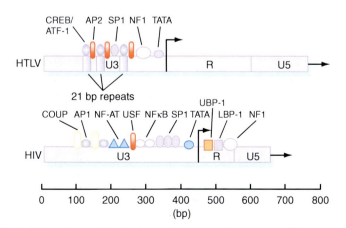

FIGURE 5.14 Cellular transcription factors that interact with retrovirus LTRs.
Many cellular DNA-binding proteins are involved in regulating both the basal and *trans*-activated levels of transcription from the promoter in the U3 region of the retrovirus LTR.

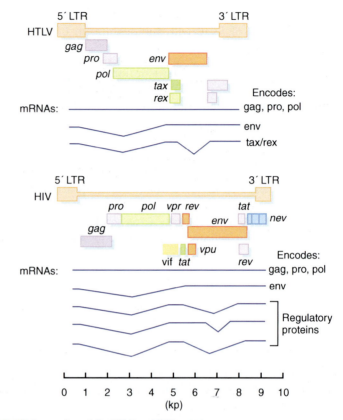

FIGURE 5.15 Expression of the HTLV and HIV genomes.
These complex retroviruses contain additional genes to the usual retrovirus pattern of gag, pol, and env, and these are expressed via a complicated mixture of spliced mRNAs.

they themselves activate (Figure 5.16). On its own, this would be an unsustainable system because it would result in unregulated positive feedback, which might be acceptable in a **lytic** replication cycle but would not be appropriate for a retrovirus integrated into the genome of the host cell. Therefore, each of these viruses encodes an additional protein (the Rex and Rev proteins in HTLV and HIV, respectively), which further regulates gene expression at a posttranscriptional level (see "Posttranscriptional Control of Expression").

Control of transcription is a critical step in virus replication and in all cases is closely regulated. Even some of the simplest virus **genomes**, such as SV40, encode proteins that regulate their transcription. Many virus genomes encode *trans*-acting factors that modify and/or direct the cellular transcription apparatus. Examples of this include HTLV and HIV, as described earlier, but also the X protein of hepadnaviruses, Rep protein of parvoviruses, E1A

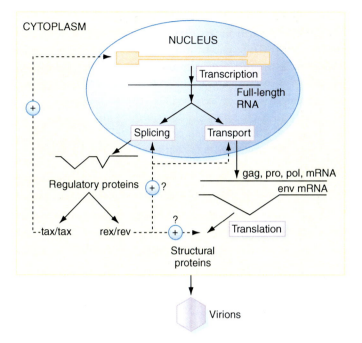

FIGURE 5.16 *Trans*-acting regulation of HTLV and HIV gene expression by virus-encoded proteins.
The Tax (HTLV) and Tat (HIV) proteins act at a transcriptional level and stimulate the expression of all virus genes. The Rex (HTLV) and Rev (HIV) proteins act posttranscriptionally and regulate the balance of expression between virion proteins and regulatory proteins.

protein of adenoviruses (see below), and the IE proteins of herpesviruses. The expression of RNA virus genomes is similarly tightly controlled, but this process is carried out by virus-encoded transcriptases and has been less intensively studied and is generally much less well understood than transcription of DNA genomes.

POSTTRANSCRIPTIONAL CONTROL OF EXPRESSION

In addition to control of the process of transcription, the expression of virus genetic information is also governed at a number of additional stages between the formation of the primary RNA transcript and completion of the finished polypeptide. Many generalized subtle controls, such as the differential stability of various mRNAs, are employed by viruses to regulate the flow of genetic information from their genomes into proteins. This section describes only a few well-researched, specific examples of posttranscriptional regulation.

Many DNA viruses that replicate in the nucleus encode mRNAs that must be spliced by cellular mechanisms to remove intervening sequences (**introns**) before being translated. This type of modification applies only to viruses that replicate in the nucleus (and not, e.g., to poxviruses) as it requires the processing of mRNAs by nuclear apparatus before they are transported into the cytoplasm for translation. However, several virus families have taken advantage of this capacity of their host cells to compress more genetic information into their genomes. A good example of such a reliance on **splicing** are the parvoviruses, transcription of which results in multiple spliced, polyadenylated transcripts in the cytoplasm of infected cells, enabling them to produce multiple proteins from their 5 kb genomes (Figure 5.5), and, similarly, polyomaviruses such as SV40 (Figure 5.12). In contrast, the large genetic capacity of herpesviruses makes it possible for these viruses to produce mostly unspliced **monocistronic** mRNAs, each of which is expressed from its own **promoter**, thereby rendering unnecessary extensive splicing to produce the required range of proteins.

One of the best-studied examples of the **splicing** of virus mRNAs is the expression of the adenovirus **genome** (Figure 5.17). Several "families" of adenovirus genes are expressed via differential splicing of precursor **hnRNA** transcripts. This is particularly true for the early genes that encode *trans-acting* regulatory proteins expressed immediately after infection. The first proteins to be expressed, E1A and E1B, are encoded by a transcriptional unit on the *r*-strand at the extreme left-hand end of the adenovirus genome

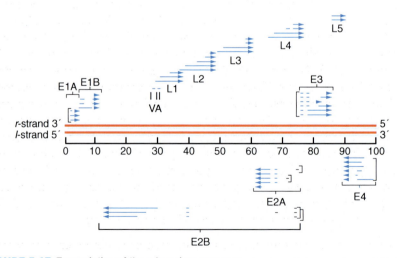

FIGURE 5.17 Transcription of the adenovirus genome.
The arrows in this figure show the position of exons in the virus genome which are joined by splicing to produce families of related but unique virus proteins.

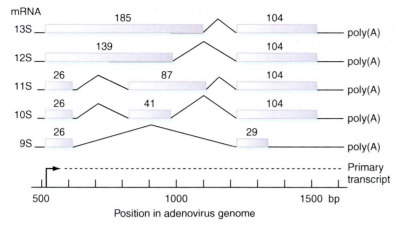

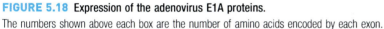

FIGURE 5.18 Expression of the adenovirus E1A proteins.
The numbers shown above each box are the number of amino acids encoded by each exon.

(Figure 5.17). These proteins are primarily transcriptional *trans*-regulatory proteins comparable to the Tax and Tat proteins described above but are also involved in **transformation** of adenovirus-infected cells (Chapter 6). Five polyadenylated, spliced mRNAs are produced (13S, 12S, 11S, 10S, and 9S) which encode five related E1A polypeptides (containing 289, 243, 217, 171, and 55 amino acids, respectively) (Figure 5.18). All of these proteins are translated from the same reading frame and have the same amino and carboxy termini. The differences between them are a consequence of differential splicing of the E1A transcriptional unit and result in major differences in their functions. The 289 and 243 amino acid peptides are transcriptional activators. Although these proteins activate transcription from all the early adenovirus promoters, it has been discovered that they also seem to be "promiscuous," activating most RNA polymerase II-responsive promoters that contain a TATA box. There are no obvious common sequences present in all of these promoters, and there is no evidence that the E1A proteins bind directly to DNA. E1A proteins from different adenovirus serotypes contain three conserved domains: CR1, CR2, and CR3. The E1A proteins interact with many other cellular proteins, primarily through binding to the three conserved domains. By binding to components of the basal transcription machinery—activating proteins that bind to upstream promoter and enhancer sequences and regulatory proteins that control the activity of DNA-binding factors—E1A can both activate and repress transcription.

Synthesis of the adenovirus E1A starts a cascade of transcriptional activation by turning on transcription of the other adenovirus early genes: E1B, E2, E3, and E4 (Figure 5.17). After the virus **genome** has been replicated, this cascade eventually results in transcription of the late genes encoding the structural

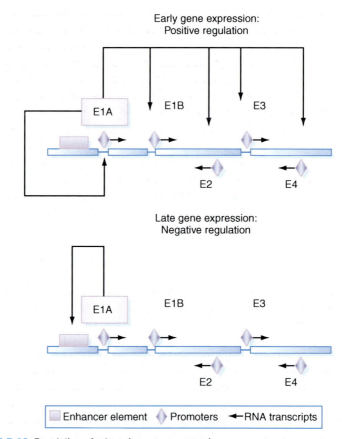

Early gene expression:
Positive regulation

E1A

E1B   E3

E2   E4

Late gene expression:
Negative regulation

E1A

E1B   E3

E2   E4

Enhancer element   ◆ Promoters   ◄─RNA transcripts

FIGURE 5.19 Regulation of adenovirus gene expression.
Adenovirus early proteins are involved in complex positive and negative regulation of gene expression.

proteins. The transcription of the E1A itself is a balanced, self-regulating system. The IE genes of DNA viruses typically have strong **enhancer elements** upstream of their promoters. This is because in a newly infected cell there are no virus proteins present and the enhancer is required to "kick-start" expression of the virus genome. The IE proteins synthesized are transcriptional activators that turn on expression of other virus genes, and E1A functions in exactly this way. However, although E1A *trans*-activates its own promoter, the protein represses the function of the upstream enhancer element so, at high concentrations, it also downregulates its own expression (Figure 5.19).

The next stage at which expression can be regulated is during export of mRNA from the nucleus and preferential translation in the cytoplasm. Again, the best-studied example of this phenomenon comes from the *Adenoviridae*. The VA genes encode two small (~160 nt) RNAs transcribed from the *r*-strand of the genome by RNA polymerase III (whose normal function is to transcribe similar small RNAs such as 5S ribosomal RNA and tRNAs) during the late phase of virus replication (Figure 5.17). Both VA RNA I and VA RNA II have a high degree of secondary structure, and neither molecule encodes any polypeptide—in these two respects they are similar to tRNAs—and they accumulate to high levels in the cytoplasm of adenovirus-infected cells. The way in which these two RNAs act is not completely understood, but their net effect is to boost the synthesis of adenovirus late proteins. The VA RNAs are processed by the host cell to form virus-encoded miRNAs. These operate through RNA interference (see Chapter 6) to downregulate a large number of cellular genes involved in RNA binding, splicing, and translation. In addition, virus infection of cells stimulates the production of interferons (Chapter 6). One of the actions of interferons is to activate a cellular protein kinase known as PKR that inhibits the initiation of translation. VA RNA I binds to this kinase, preventing its activity and relieving inhibition on translation. The effects of interferons on the cell are generalized (discussed in Chapter 6) and result in inhibition of the translation of both cellular and virus mRNAs. The effect of the VA RNAs is to promote selectively the translation of adenovirus mRNAs at the expense of cellular mRNAs whose translation is inhibited.

The HTLV Rex and HIV Rev proteins mentioned earlier also act to promote the selective translation of specific virus mRNAs. These proteins regulate the differential expression of the virus genome but do not substantially alter the expression of cellular mRNAs. Both of these proteins appear to function in a similar way, and, although not related to one another in terms of their amino acid sequences, the HTLV Rex protein can substitute functionally for the HIV Rev protein. Negative-regulatory sequences in the HIV and HTLV genomes cause the retention of virus mRNAs in the nucleus of the infected cell. These sequences are located in the intron regions that are removed from spliced mRNAs encoding the Tax/Tat and Rex/Rev proteins (Figure 5.15), therefore, these proteins are expressed immediately after infection. Tax and Tat stimulate enhanced transcription from the virus LTR (Figure 5.16). However, unspliced or singly spliced mRNAs encoding the *gag, pol,* and *env* gene products are only expressed when sufficient Rex/Rev protein is present in the cell. Both proteins bind to a region of secondary structure formed by a particular sequence in the mRNA and shuttle between the nucleus and the cytoplasm as they contain both a nuclear localization signal and a nuclear export signal, increasing the export of unspliced virus mRNA to the cytoplasm where it is translated and acts as the virus genome during particle formation.

The efficiency with which different mRNAs are translated varies considerably and is determined by a number of factors, including the stability and secondary structure of the RNA, but the main one appears to be the particular nucleotide sequence surrounding the AUG translation initiation codon that is recognized by ribosomes. The most favorable sequence for initiation is GCC(A/G)CCAUGGG, although there can be considerable variation within this sequence. A number of viruses use variations of this sequence to regulate the amounts of protein synthesized from a single mRNA. Examples include the Tax and Rex proteins of HTLV which are encoded by overlapping reading frames in the same doubly spliced 2.1 kb mRNA (Figure 5.15). The AUG initiation codon for the Rex protein is upstream of that for Tax but provides a less favorable context for initiation of translation than the sequence surrounding the Tax AUG codon. This is known as the "leaky scanning" mechanism because it is believed that the ribosomes scan along the mRNA before initiating translation. Therefore, the relative abundance of Rex protein in HTLV-infected cells is considerably less than that of the Tax protein, even though both are encoded by the same mRNA.

Picornavirus genomes illustrate an alternative mechanism for controlling the initiation of translation. Although these genomes are genetically economical (i.e., have discarded most cis-acting control elements and express their entire coding capacity as a single polyprotein), they have retained long NCRs at their 5′ ends, comprising approximately 10% of the entire genome. These sequences are involved in the replication and possibly packaging of the virus genome. Translation of most cellular mRNAs is initiated when ribosomes recognize the 5′ end of the mRNA and scan along the nucleotide sequence until they reach an AUG initiation codon. Picornavirus genomes are not translated in this way. The 5′ end of the RNA is not capped and thus is not recognized by ribosomes in the same way as other mRNAs, but it is modified by the addition of the VPg protein (see Chapters 3 and 6). There are also multiple AUG codons in the 5′ NCR upstream of the start of the polyprotein coding sequences which are not recognized by ribosomes. In picornavirus-infected cells, a virus protease cleaves the 220-kDa "cap-binding complex" involved in binding the m7G cap structure at the 5′ end of the mRNA during initiation of translation. Translation of artificially mutated picornavirus mRNAs in vitro and the construction of bicistronic picornavirus genomes bearing additional 5′ NCR signals in the middle of the polyprotein have resulted in the concept of the ribosome "landing pad," or internal ribosomal entry site (IRES). Rather than scanning along the RNA from the 5′ end, ribosomes bind to the RNA via the IRES and begin translation internally. This is a precise method for controlling the translation of virus proteins. Very few cellular mRNAs utilize this mechanism but it has been shown to be used by a variety of viruses, including picornaviruses, hepatitis C virus, coronaviruses, and flaviviruses.

Many viruses belonging to different families compress their genetic information by encoding different polypeptides in overlapping reading frames. The problem with this strategy lies in decoding the information. If each polypeptide is expressed from a monocistronic mRNA transcribed from its own promoter, the additional *cis*-acting sequences required to control and coordinate expression might cancel out any genetic advantage gained. More importantly, there is the problem of coordinately regulating the transcription and translation of multiple different messages. Therefore, it is highly desirable to express several polypeptides from a single RNA transcript, and the examples described above illustrate several mechanisms by which this can be achieved, e.g., differential splicing and control of RNA export from the nucleus, or initiation of translation.

BOX 5.4 VIRUSES—MAKING MORE FROM LESS

If viruses were as wasteful with their genomes as cells are, they would struggle to exist. For most cellular genes, it's one gene, one protein. And while some of the big DNA viruses do that, most viruses have to work harder and get more than one protein out of a gene. There are lots of ways they do this. Splicing, alternative start codons, ribosomal frameshifting, alternate protease cleavages—all these are used by different viruses to squeeze the maximum amount of information into the minimum space. Of course the trick is to ensure that you can get the information, in the form of proteins, out again. The host cell is tricked into doing this, so in addition to the range of virus genomes, the range of host organisms also means that many different gene expression mechanisms are used.

An additional mechanism known as "ribosomal frameshifting" is used by several groups of viruses to achieve the same effect. The best-studied examples of this phenomenon come from retrovirus genomes, but many viruses use a similar mechanism. Such frameshifting was first discovered in viruses but is now known to occur also in prokaryotic and eukaryotic cells. Retrovirus genomes are transcribed to produce at least two 5′ capped, 3′ polyadenylated mRNAs. Spliced mRNAs encode the envelope proteins, as well as, in more complex retroviruses such as HTLV and HIV, additional proteins such as the Tax/Tat and Rex/Rev proteins (Figure 5.15). A long, unspliced transcript encodes the *gag*, *pro*, and *pol* genes and also forms the genomic RNA packaged into virions. The problem faced by retroviruses is how to express three different proteins from one long transcript. The arrangement of the three genes varies in different viruses. In some cases (e.g., HTLV), they occupy three different reading frames, while in others (e.g., HIV), the protease (*pro*) gene forms an extension at the 5′ end of the *pol* gene (Figure 5.20). In the latter case, the protease and polymerase (i.e., reverse transcriptase) are expressed as a polyprotein that is autocatalytically cleaved into the mature proteins in a process that is similar to the cleavage of picornavirus polyproteins.

At the boundary between each of the three genes is a particular sequence that usually consists of a tract of reiterated nucleotides, such as UUUAAAC (Figure 5.21). This sequence is rarely found in protein-coding sequences and

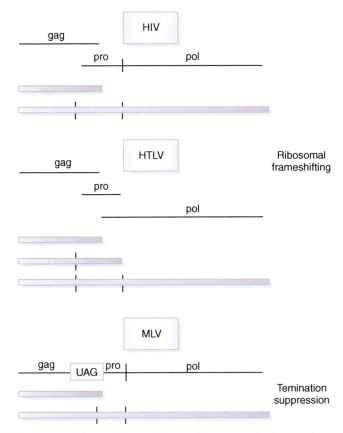

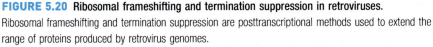

FIGURE 5.20 Ribosomal frameshifting and termination suppression in retroviruses.
Ribosomal frameshifting and termination suppression are posttranscriptional methods used to extend the range of proteins produced by retrovirus genomes.

therefore appears to be specifically used for this type of regulation. Most ribosomes encountering this sequence will translate it without difficulty and continue on along the transcript until a translation stop codon is reached. However, a proportion of the ribosomes that attempt to translate this sequence will slip back by one nucleotide before continuing to translate the message, but now in a different (i.e., −1) reading frame. Because of this, the UUUAAAC sequence has been termed the "slippery sequence," and the result of this frameshifting is the translation of a polyprotein containing alternative information from a different reading frame. This mechanism also allows the virus to control the ratios of the proteins produced. Because only a proportion of ribosomes undergoes frameshifting at each slippery sequence, there is a gradient of translation from the reading frames at the 5′ end of the mRNA to those at the 3′ end.

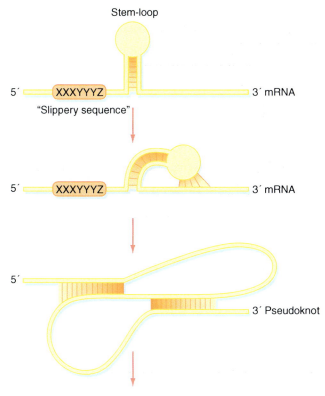

Stem-loop

5′ ▭ XXXYYYZ ▭ 3′ mRNA

"Slippery sequence"

5′ ▭ XXXYYYZ ▭ 3′ mRNA

5′

3′ Pseudoknot

FIGURE 5.21 RNA pseudoknot formation.
RNA pseudoknot formation is the mechanism by which ribosomal frameshifting occurs in a number of different viruses and a few cellular genes (see text for details).

The slippery sequence alone results in only a low frequency of frameshifting, which appears to be inadequate to produce the amount of protease and reverse transcriptase protein required by the virus. So there are additional sequences that further regulate this system and increase the frequency of frameshift events. A short distance downstream of the slippery sequence is an inverted repeat that allows the formation of a stem−loop structure in the mRNA (Figure 5.21). A little further on is an additional sequence complementary to the nucleotides in the loop that allows base-pairing between these two regions of the RNA. The net result of this combination of sequences is the formation of what is known as an RNA pseudoknot. This secondary structure in the mRNA causes ribosomes translating the message to pause at the position of the slippery sequence upstream, and this slowing or pausing of the ribosome during translation increases the frequency at which frameshifting occurs, thus boosting the relative amounts of the proteins encoded by the downstream reading frames. It is easy to imagine how this system can be fine-tuned by subtle mutations that alter the stability of the pseudoknot structure and thus the relative expression of the different genes.

Yet another method of translational control is termination suppression. This is a mechanism similar in many respects to frameshifting that permits multiple polypeptides to be expressed from individual reading frames in a single mRNA. In some retroviruses, such as murine leukemia virus (MLV), the *pro* gene is separated from the *gag* gene by a UAG termination codon rather than a slippery sequence and pseudoknot (Figure 5.20). In the majority of cases, translation of MLV mRNA terminates at this sequence, giving rise to the Gag proteins. However in a few instances, the UAG stop codon is suppressed and translation continues, producing a Gag–Pro–Pol polyprotein, which subsequently cleaves itself to produce the mature proteins. The overall effect of this system is much the same as ribosomal frameshifting, with the relative ratios of Gag and Pro/Pol proteins being controlled by the frequency with which ribosomes traverse or terminate at the UAG stop codon.

SUMMARY

Control of gene expression is a vital element of virus replication. Coordinate expression of groups of virus genes results in successive phases of gene expression. Typically, IE genes encode "activator" proteins, early genes encode further regulatory proteins, and late genes encode virus structural proteins. Viruses make use of the biochemical apparatus of their host cells to express their genetic information as proteins and, consequently, utilize the appropriate biochemical language recognized by the cell. Thus viruses of prokaryotes produce polycistronic mRNAs, while viruses with eukaryotic hosts produce more monocistronic mRNAs. Some viruses of eukaryotes do produce polycistronic mRNA to assist with the coordinate regulation of multiple genes.

Further Reading

Alberts, B. (Ed.), 2014. Molecular Biology of the Cell. fifth ed. Garland Science, New York, ISBN 0815344643.

Firth, A.E., Brierley, I., 2012. Non-canonical translation in RNA viruses. J. Gen. Virol. 93 (7), 1385–1409.

Gómez-Díaz, E., Jordà, M., Peinado, M.A., Rivero, A., 2012. Epigenetics of host–pathogen interactions: the road ahead and the road behind. PLoS Pathog. 8 (11), e1003007.

Kannian, P., Green, P.L., 2010. Human T lymphotropic virus type 1 (HTLV-1): molecular biology and oncogenesis. Viruses 2 (9), 2037–2077.

Karn, J., Stoltzfus, C.M., 2012. Transcriptional and posttranscriptional regulation of HIV-1 gene expression. Cold Spring Harb. Perspect. Med. 2 (2), a006916.

Kincaid, R.P., Sullivan, C.S., 2012. Virus-encoded microRNAs: an overview and a look to the future. PLoS Pathog. 8 (12), e1003018.

López-Lastra, M., et al., 2010. Translation initiation of viral mRNAs. Rev. Med. Virol. 20 (3), 177−195.

Ptashne, M., 2004. A Genetic Switch: Phage Lambda Revisited. Cold Spring Harbor Laboratory Press, Cold Spring Harbor, New York, ISBN 0879697164.

Resa-Infante, P., Jorba, N., Coloma, R., Ortín, J., 2011. The influenza RNA synthesis machine. RNA Biol. 8 (2), 207−215.

Santos, F., Martinez-Garcia, M., Parro, V., Antón, J., 2014. Microarray tools to unveil viral-microbe interactions in nature. Front. Ecol. Evol. 2, 31.

Skalsky, R.L., Cullen, B.R., 2010. Viruses, microRNAs, and Host interactions. Annu. Rev. Microbiol. 64, 123−141.

Zhao, H., Dahlö, M., Isaksson, A., Syvänen, A.C., Pettersson, U., 2012. The transcriptome of the adenovirus infected cell. Virology 424 (2), 115−128.

CHAPTER 6

Infection

Intended Learning Outcomes

On completing this chapter you should be able to:

- Discuss the similarities and the differences between virus infections of plants and of animals.
- Explain how the immune responses to viruses enables the body to resist infection, and how viruses respond to this pressure.
- Describe and understand how virus infections are prevented and treated.

VIRUS INFECTIONS OF PLANTS

Life on Earth depends on the primary productivity of plants—the production of organic molecules from inorganic molecules such as CO_2—(with some an additional contribution from some bacteria). From the smallest single-celled alga in the ocean to the largest forest giant tree (and everything in between, such as broccoli), they are vitally important. Photosynthetic algae in the oceans play a major role in controlling the atmosphere and the climate, and interaction with viruses is one of the major mechanisms which in turn control the algae. All higher animals depend on the primary productivity of plants for their food. So plants are a big deal, and anything which affects plant growth is of great importance.

In purely economic terms, viruses are only of importance if it is likely that they will affect crops during their commercial lifetime, a likelihood that varies greatly between very short extremes in horticultural production and very long extremes in forestry. Some estimates have put total worldwide cost of plant virus infections as high as US\$ 6×10^{10} per year. The mechanism

CONTENTS

173

Principles of Molecular Virology. DOI: http://dx.doi.org/10.1016/B978-0-12-801946-7.00006-7

BOX 6.1: IS BOTANY BORING?

Some of the most original experimental biology currently being done involves plant science. Biologists can do experiments with plants that they can only dream of being able to perform with animals. And yet the idea persists among many that botany is boring. Much of the most exciting plant science involves plant viruses, either as experimental tools or in terms of finding ways to prevent infection. And as this section describes, the biology of plant viruses has some striking differences from that of animal viruses. So if you think botany is boring, you probably need to find out more about plant viruses.

by which plant viruses are transmitted between hosts is therefore of great importance. There are a number of routes by which plant viruses may be transmitted:

- **Seeds:** These may transmit virus infection either by external contamination of the seed with virus particles or by infection of the living tissues of the embryo. Transmission by this route leads to early outbreaks of disease in new crops which are usually initially focal in distribution but may subsequently be transmitted to the remainder of the crop by other mechanisms.
- **Vegetative propagation/grafting:** These techniques are inexpensive and easy methods of plant propagation but provide the ideal opportunity for viruses to spread to new plants.
- **Vectors:** Many different groups of living organisms can act as vectors and spread viruses from one plant to another:
  - Bacteria (e.g., *Agrobacterium tumefaciens*—the Ti plasmid of this organism has been used experimentally to transmit virus genomes between plants)
  - Fungi
  - Nematodes
  - Arthropods: insects (e.g., aphids, leafhoppers, planthoppers, beetles, thrips)
  - Arachnids (e.g., mites)
- **Mechanical:** Mechanical transmission of viruses is the most widely used method for experimental infection of plants and is usually achieved by rubbing virus-containing preparations into the leaves, which in most plant species are particularly susceptible to infection. However, this is also an important natural method of transmission. Virus particles may contaminate soil for long periods and be transmitted to the leaves of new host plants as wind-blown dust or as rain-splashed mud.

The problems plant viruses face in initiating infections of host cells have already been described (Chapter 4), as has the fact that no known plant virus employs a specific cellular receptor of the types that animal and bacterial viruses use to attach to cells. Transmission of plant viruses by insects is of particular agricultural importance. Huge areas of monoculture and the inappropriate use of pesticides that kill natural predators can result in massive population booms of pest insects such as aphids. Plant viruses rely on a mechanical breach of the integrity of a cell wall to directly introduce a virus particle into a cell. This is achieved either by the vector associated with transmission of the virus or simply by mechanical damage to cells. Transfer by insect vectors is a particularly efficient means of virus transmission. In some instances, viruses are transmitted mechanically from one plant to the next by the vector and the insect is only a means of distribution, through flying or being carried on the wind for long distances (sometimes hundreds of miles). Insects that bite or suck plant tissues are the ideal means of transmitting viruses to new hosts—a process known as nonpropagative transmission. However, in other cases (e.g., many plant rhabdoviruses), the virus may also infect and multiply in the tissues of the insect (propagative transmission) as well as those of host plants. In these cases, the vector serves as a means not only of distributing the virus but also of amplifying the infection.

Initially, most plant viruses multiply at the site of infection, giving rise to localized symptoms such as necrotic spots on the leaves. The virus may subsequently be distributed to all parts of the plant either by direct cell-to-cell spread or by the vascular system, resulting in a systemic infection involving the whole plant. However, the problem these viruses face in reinfection and recruitment of new cells is the same as the one they faced initially—how to cross the barrier of the plant cell wall. Plant cell walls necessarily contain channels called "plasmodesmata" which allow plant cells to communicate with each other and to pass metabolites between them. However, these channels are too small to allow the passage of virus particles or genomic nucleic acids. Many (if not most) plant viruses have evolved specialized movement proteins that modify the plasmodesmata. One of the best known examples of this is the 30-k protein of tobacco mosaic virus (TMV). This protein is expressed from a subgenomic mRNA (Figure 3.12), and its function is to modify plasmodesmata causing genomic RNA coated with 30-k protein to be transported from the infected cell to neighboring cells (Figure 6.1). Other viruses, such as cowpea mosaic virus (CPMV; *Comoviridae*) have a similar strategy but employ a different molecular mechanism. In CPMV, the 58-/48-k proteins form tubular structures allowing the passage of intact virus particles to pass from one cell to another (Figure 6.1).

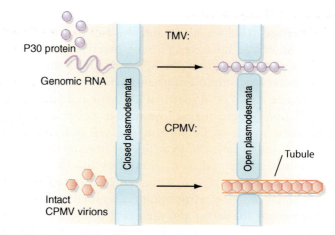

FIGURE 6.1 Plant movement proteins.

Plant movement proteins allow plant viruses to infect new cells without having to penetrate the cell wall from the outside for each new cell.

Typically, virus infections of plants might result in effects such as growth retardation, distortion, mosaic patterning on the leaves, yellowing, wilting, etc. These macroscopic symptoms result from:

- Necrosis of cells, caused by direct damage due to virus replication,
- Hypoplasia—localized retarded growth frequently leading to mosaicism (the appearance of thinner, yellow areas on the leaves),
- Hyperplasia—excessive cell division or the growth of abnormally large cells, resulting in the production of swollen or distorted areas of the plant.

Plants might be seen as sitting targets for virus infection—unlike animals, they cannot run away. However, plants exhibit a sophisticated range of responses to virus infections designed to minimize harmful effects. Plants fight virus infections in a number of ways. First, they need to detect the infection, which they do by means of sensing virus signature molecules (so-called pathogen-associated molecular patterns or PAMPs, e.g., particular proteins) via dedicated receptors. When this happens, the production of resistance proteins that activate highly specific resistance mechanisms is triggered. In response, plant viruses attempt to evade these defense mechanisms by altering protein structures where possible and by producing proteins which bind to and hide small RNAs which would trigger RNA silencing. Infection results in a "hypersensitive response," manifested as:

- The synthesis of a range of new proteins, the pathogenesis-related ("PR") proteins,
- An increase in the production of cell wall phenolic substances,
- The release of active oxygen species,

- The production of phytoalexins,
- The accumulation of salicylic acid—amazingly, plants can even warn each other that viruses are coming by airborne signaling with volatile compounds such as methyl salicylate.

The hypersensitive response involves synthesis of a wide range of different molecules. Some of these PR proteins are proteases, which presumably destroy virus proteins, limiting the spread of the infection. There is some similarity here between the design of this response and the production of interferons (IFNs) by animals.

Systemic resistance to virus infection is a naturally occurring phenomenon in some strains of plant. This is clearly a highly desirable characteristic that is prized by plant breeders, who try to spread this attribute to economically valuable crop strains. There are probably many different mechanisms involved in systemic resistance, but in general terms there is a tendency of these processes to increase local necrosis when substances such as proteases and per-oxidases are produced by the plant to destroy the virus and to prevent its spread and subsequent systemic infection. An example of this is the tobacco N gene, which encodes a cytoplasmic protein with a nucleotide-binding site which interferes with the TMV replicase. When present in plants, this gene causes TMV to produce a localized, necrotic infection rather than the systemic mosaic symptoms normally seen. There are many different mechanisms involved in systemic resistance, but in general terms there is a tendency toward increased local necrosis as substances such as proteases and peroxidases are produced by the plant to destroy the virus and to prevent its spread and subsequent systemic disease.

Virus-resistant plants have been created by the production of transgenic plants expressing recombinant virus proteins or nucleic acids which interfere with virus replication without producing the pathogenic consequences of infection, for example:

- Virus coat proteins, which have a variety of complex effects, including inhibition of virus uncoating and interference of expression of the virus at the level of RNA ("gene silencing" by "untranslatable" RNAs),
- Intact or partial virus replicases which interfere with genome replication,
- Antisense RNAs,
- Defective virus genomes,
- Satellite sequences (see Chapter 8),
- Catalytic RNA sequences (ribozymes),
- Modified movement proteins.

This is a very promising technology that offers the possibility of substantial increases in agricultural production without the use of expensive, toxic, and ecologically damaging chemicals (fertilizers, herbicides, or pesticides). In some countries, notably in Europe, public resistance to genetically engineered plants has so far prevented the widespread adoption of new varieties produced by genetic manipulation without considering the environmental cost of not utilizing these new approaches to plant breeding.

IMMUNE RESPONSES TO VIRUS INFECTIONS IN ANIMALS

The most significant response to virus infection in vertebrates is activation of both the cellular and humoral parts of the immune system. A complete description of all the events involved in the immune response to the presence of foreign antigens is beyond the scope of this book, so you should refer to the books mentioned in the Further Reading at the end of this chapter to ensure that you are familiar with all the immune mechanisms (and jargon!) described below. A brief summary of some of the more important aspects is worth considering however, beginning with the humoral immune response, which results in the production of antibodies.

The major impact of the humoral immune response is the eventual clearance of virus from the body. Serum neutralization stops the spread of virus to uninfected cells and allows other defense mechanisms to mop up the infection. Figure 6.2 shows a very simplified version of the mammalian humoral response to infection. Virus infection induces at least three classes of antibody: immunoglobulin G (IgG), IgM, and IgA. IgM is a large, multivalent molecule that is most effective at cross-linking large targets (e.g., bacterial cell walls or flagella) but is probably less important in combating virus infections. In contrast, the production of IgA is very important for initial

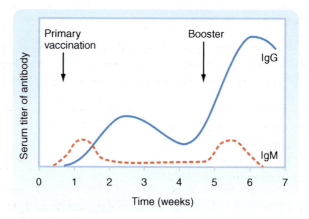

FIGURE 6.2 Kinetics of the immune response.
Simplified version of the kinetics of the mammalian humoral response to a "typical" foreign virus (or other) antigen.

protection from virus infection. Secretory IgA is produced at mucosal surfaces and results in "mucosal immunity," an important factor in preventing infection from occurring. Induction of mucosal immunity depends to a large extent on the way in which antigens are presented to and recognized by the immune system. Similar antigens incorporated into different vaccine delivery systems (see "Prevention and Therapy of Virus Infection") can lead to very different results in this respect, and mucosal immunity is such an important factor that similar vaccines may vary considerably in their efficacy. IgG is probably the most important class of antibody for direct neutralization of virus particles in serum and other body fluids (into which it diffuses).

Direct virus neutralization by antibodies results from a number of mechanisms, including conformational changes in the virus capsid caused by antibody binding, or blocking of the function of the virus target molecule (e.g., receptor binding) by steric hindrance. A secondary consequence of antibody binding is phagocytosis of antibody-coated ("opsonized") target molecules by mononuclear cells or polymorphonuclear leukocytes. This results from the presence of the Fc receptor on the surface of these cells, but as has already been noted in Chapter 4, in some cases opsonization of virus by the binding of nonneutralizing antibodies can result in enhanced virus uptake. This has been shown to occur with rabies virus, and in the case of human immunodeficiency virus (HIV) may promote uptake of the virus by macrophages. Nonphagocytic cells can also destroy antibody-coated viruses via an intracellular pathway involving the TRIM21 protein. Antibody binding also leads to the activation of the complement cascade, which assists in the neutralization of virus particles. Structural alteration of virus particles by complement binding can sometimes be visualized directly by electron microscopy. Complement is particularly important early in virus infection when limited amounts of low-affinity antibody are made—complement enhances the action of these early responses to infection.

Despite all the above mechanisms, in overall terms cell-mediated immunity is probably more important than humoral immunity in the control of virus infections. This is demonstrated by the following observations:

- Congenital defects in cell-mediated immunity tend to result in predisposition to virus (and parasitic) infections, rather than to bacterial infections.
- The functional defect in acquired immune deficiency syndrome (AIDS) is a reduction in the ratio of T-helper ($CD4^+$):T-suppressor ($CD8^+$) cells from the normal value of about 1.2 to 0.2. AIDS patients commonly suffer many opportunistic virus infections (e.g., various herpesviruses such as herpes simplex virus [HSV], cytomegalovirus [CMV], and Epstein–Barr virus [EBV]), which may have been present before the onset of AIDS but were previously suppressed by the intact immune system.

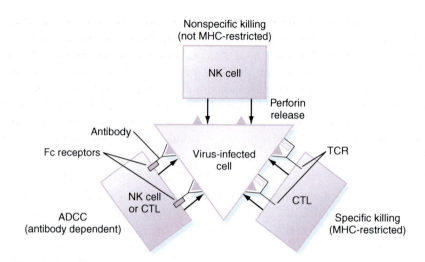

FIGURE 6.3 Mechanisms of cell-mediated immunity.
Diagram illustrating the three main mechanisms by which cell-mediated immunity kills virus-infected cells.

Cell-mediated immunity depends on three main effects (Figure 6.3). These all act via molecular mechanisms that will be explained later in this chapter (see "Viruses and Apoptosis," below):

- Nonspecific cell killing (mediated by "natural killer" [NK] cells),
- Specific cell killing (mediated by cytotoxic T-lymphocytes [CTLs]),
- Antibody-dependent cellular cytotoxicity (ADCC).

NK cells carry out cell lysis independently of conventional immunological specificity, that is, they do not depend on clonal antigen recognition for their action. They are not major histocompatibility complex (MHC) restricted. In other works, NK cells are able to recognize virus-infected cells without being presented with a specific antigen by a macromolecular complex consisting of MHC antigens plus the T-cell receptor/CD3 complex. The advantage of this is that NK cells have broad specificity (many antigens rather than a single epitope) and are also active without the requirement for sensitizing antibodies. They are therefore the first line of defense against virus infection. NK cells are most active in the early stages of infection (i.e., in the first few days), and their activity is stimulated by IFN-α/β. NK cells are not directly induced by virus infection—they exist even in immunologically naive individuals and are "revealed" in the presence of IFN-α/β. They are thus part of the "innate" rather than the "adaptive" immune response. Their function is complementary to and is later taken over by CTLs which are part of the "adaptive" immune response. Not all of the targets for NK cells on the surface of infected cells are known, but they are inhibited by MHC class I antigens (which are present on all nucleated cells), allowing recognition of "self" (i.e., uninfected cells) and preventing total destruction of the body. It is well known that some virus infections disturb normal cellular MHC-I expression and this is one mechanism by which NK cells recognize virus-infected cells. NK cell cytotoxicity is activated by IFN-α/β, directly linking NK cell activity to virus infection.

Unlike NK cells which may be either CD4$^+$ or CD8$^+$, CTLs are usually of CD8$^+$ (suppressor) phenotype, that is, they express CD8 molecules on their surface. CTLs are the major cell-mediated immune response to virus infections and are MHC restricted—clones of cells recognize a specific antigen only when presented by MHC-I antigen on the target cell to the T-cell receptor/CD3 complex on the surface of the CTL. (MHC-I antigens are expressed on all nucleated cells in the body; MHC class II antigens are expressed only on the surface of the antigen-presenting cells of the immune system—T-cells, B-cells, and macrophages.) CTL activity requires "help" (i.e., cytokine production) from T-helper cells. The CTLs themselves recognize foreign antigens through the T-cell receptor/CD3 complex, which "docks" with antigen presented by MHC-I on the surface of the target cell (Figure 6.4). The mechanism of cell killing by CTL is similar to that of NK cells (explained below). The induction of a CTL response also results in the release of many different cytokines from T-helper cells, some of which result in clonal proliferation of antigen-specific CTL and others that have direct antiviral effects—for example, IFNs. The kinetics of the CTL response (peaking at about 7 days after infection) is somewhat slower than the NK response (e.g., 3−7 days cf. 0.5−3 days)—so NK cells and CTLs are complementary systems.

The induction of a CTL response is dependent on recognition of specific T-cell epitopes by the immune system. These are distinct from the B-cell epitopes recognized by the humoral arm of the immune system. T-cell epitopes are more highly conserved (less variable) than B-cell epitopes, which are more able to mutate quickly to escape immune pressure. These are important considerations in the design of antiviral vaccines. The specificity of cell killing by CTLs is not absolute. Although they are better "behaved" than NK cells, diffusion of perforin and local cytokine production frequently results in inflammation and bystander cell damage. This is a contributory cause of the pathology of many virus diseases (see Chapter 7), but the less attractive alternative is to allow virus replication to proceed unchecked.

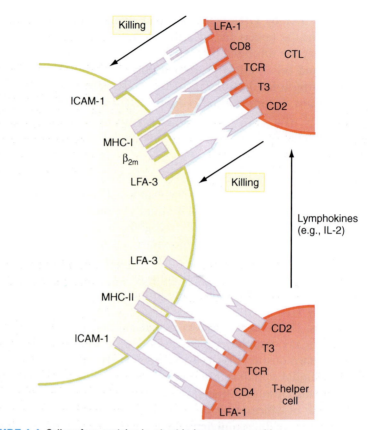

FIGURE 6.4 Cell-surface proteins involved in immune recognition.
Close contact between cells results in cell-to-cell signaling which regulates the immune response.

ADCC is less well understood than either of the two mechanisms mentioned above. ADCC can be carried out by NK cells or by CTLs. The mechanism of cell killing is the same as that described in the next section, although complement may also be involved in ADCC. The distinguishing feature of ADCC is that this mechanism is dependent on the recognition of antigen on the surface of the target cell by means of antibody on the surface of the effector cell. The antibody involved is usually IgG, which is bound to Fc receptors on the surface of the T-cell. ADCC therefore requires a preexisting antibody response and hence does not occur early during primary virus infections—it is part of the adaptive immune response. The overall contribution of ADCC to the control of virus infections is not clear, although it is now believed that it plays a significant part in their control.

BOX 6.2 COLLATERAL DAMAGE

We all walk around with a time bomb inside us. It's called your immune system. When it ticks away quietly in the background, we don't notice it, but when things go wrong ... it's very bad news. Your immune system has to keep working with Goldilocks precision—not too strong, not too weak—for decade after decade. And as soon as a virus turns up and starts to take over your cells, your immune system has to show up right away (leave it a few days and it's probably too late), and it has to get it right every time. Fighting viruses is warfare and people get hurt—mostly you. Fever, muscle pain, headaches, vomiting, dead neurons in your brain or spinal cord. That's all down to your immune system. But maybe you'd prefer encephalitis?

VIRUSES AND APOPTOSIS

Apoptosis, or "programmed cell death," is a critical mechanism in tissue remodeling during development and in cell killing by the immune system. There are two ways in which a cell can die: necrosis or apoptosis.

- Necrosis is the normal response of cells to injury caused by toxins or environmental stress. Necrosis is marked by nonspecific changes such as disruption of the plasma membrane and nuclear envelope, rupture of membrane-bounded organelles such as mitochondria and lysosomes, cell swelling, random fragmentation of DNA/RNA, influx of calcium ions into the cell, and loss of membrane electrical potential. The release of cellular components from the dying cell causes a localized inflammatory response by the cells of the immune system. This frequently leads to damage to adjacent cells/tissue—"bystander" cell damage.
- Apoptosis is, in contrast, a tightly regulated process that relies on complex molecular cascades for its control. It is marked by cell shrinkage, condensation, and clumping of chromatin, a regular pattern of DNA fragmentation, and "bubbling off" of cellular contents into small membrane-bounded vesicles ("blebbing") which are subsequently phagocytosed by macrophages, preventing inflammation.

When triggered by the appropriate signals, immune effector cells such as CTLs and NK cells release previously manufactured lytic granules stored in their cytoplasm. These act on the target cell and induce apoptosis by two mechanisms:

- Release of cytotoxins such as: (1) perforin (aka cytolysin), a peptide related to complement component C9 which, on release, polymerizes to form polyperforin, which forms transmembrane channels, resulting in permeability of the target cell membrane; and (2) granzymes, which are

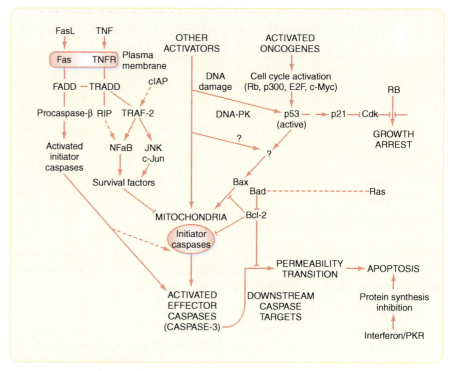

FIGURE 6.5 Overview of apoptosis.
The pathways controlling apoptosis are very complex. This diagram represents only a simple summary of some of the mechanisms of major significance in virus infections.

serine proteases related to trypsin. These two effectors act collaboratively, the membrane pores allowing the entry of granzymes into the target cell. The membrane channels also allow the release of intracellular calcium from the target cell, which also acts to trigger apoptotic pathways.

■ In addition, CTLs (but not NK cells) express Fas ligand on their surface which binds to Fas on the surface of the target cell, triggering apoptosis. Binding of Fas ligand on the effector cell to Fas (CD95) on the target cell results in activation of cellular proteases known as "caspases," which in turn trigger a cascade of events leading to apoptosis.

The process of induction and repression of apoptosis during virus infection has received much attention during the last few years. It is now recognized that this is an important innate response to virus infection. The regulation of apoptosis is a complex issue that cannot be described fully here (see Further Reading and Figure 6.5 for a summary), but virus infections disturb normal cellular biochemistry and frequently trigger an apoptotic response, for example:

■ **Receptor signaling**: Binding of virus particles to cellular receptors may also trigger signaling mechanisms resulting in apoptosis (e.g., HIV [see Chapter 7], reovirus).

- **PKR activation:** The IFN effector PKR (RNA-activated protein kinase may be activated by some viruses (e.g., HIV, reovirus).
- **p53 activation:** Viruses that interact with p53 (Chapter 7) may cause either growth arrest or apoptosis (e.g., adenoviruses, SV40, papillomaviruses).
- **Transcriptional disregulation:** Viruses that encode transcriptional regulatory proteins may trigger an apoptotic response (e.g., HTLV Tax).
- **Foreign protein expression:** Overexpression of virus proteins at late stages of the replication cycle can also cause apoptosis by a variety of mechanisms.

In response to this cellular alarm system, many if not most viruses have evolved mechanisms to counteract this effect and repress apoptosis:

- **Bcl-2 homologues:** A number of viruses encode Bcl-2 (a negative regulator of apoptosis) homologues (e.g., adenovirus E1B-19k, human herpesvirus 8 [HHV-8] KSbcl-2).
- **Caspase inhibition:** Caspases are a family of cysteine proteases that are important inducers of apoptosis. Inhibiting these enzymes is an effective way of preventing apoptosis (e.g., baculovirus p35, serpins, vIAPs—"inhibitors of apoptosis").
- **Fas/TNF inhibition:** Viruses have evolved several mechanisms to block the effects of Fas/TNF, including blocking signaling through the plasma membrane (e.g., adenovirus E3), tumor necrosis factor receptor (TNFR) mimics (e.g., poxvirus crmA), mimics of death signaling factors (vFLIPs), and interactions with signaling factors such as Fas-associated death domain (FADD) and TNFR-associated death domain (TRADD) (e.g., HHV-4 [EBV] LMP-1).
- **p53 inhibition:** A number of viruses that interact with p53 have evolved proteins to counteract possible triggering of apoptosis (e.g., adenovirus E1B-55k and E4, SV40 T-antigen, papillomavirus E6).
- **Miscellaneous:** Many other apoptosis-avoidance mechanisms have been described in a wide variety of viruses.

Without such inhibitory mechanisms, most viruses would simply not be able to replicate due to the death of the host cell before the replication cycle was complete. However, there is evidence that at least some viruses use apoptosis to their benefit. Positive-sense RNA viruses such as poliovirus, hepatitis A virus, and Sindbis virus with lytic replication cycles appear to be able to regulate apoptosis, initially repressing it to allow replication to take place, then inducing it to allow the release of virus particles from the cell.

INTERFERONS

By the 1950s, interference (i.e., the blocking of a virus infection by a competing virus) was a well-known phenomenon in virology. In some cases, the

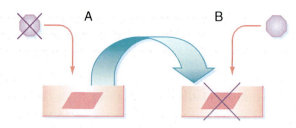

FIGURE 6.6 Discovery of IFNs.

In 1957, Alick Issacs and Jean Lindenmann discovered IFNs by performing the following experiment. (A) Pieces of chick chorioallantoic membrane were exposed to UV-inactivated (noninfectious) influenza virus in tissue culture. (B) The "conditioned" medium from these experiments (which did not contain infectious virus) was found to inhibit the infection of fresh pieces of chick chorioallantoic membrane by (infectious) influenza virus in separate cultures. They called inhibitory substance in the condition medium "interferon."

mechanism responsible is quite simple. For example, avian retroviruses are grouped into nine interference groups (A through I), based on their ability to infect various strains of chickens, pheasants, partridges, quail, etc., or cell lines derived from these species. In this case, the inability of particular viruses to infect the cells of some strains is due to the expression of the envelope glycoprotein of an endogenous provirus present in the cells which sequesters the cellular receptor needed by the exogenous virus for infection. In other cases, the mechanism of virus interference is less clear.

In 1957, Alick Issacs and Jean Lindenmann were studying this phenomenon and performed the following experiment. Pieces of chick chorioallantoic membrane were exposed to ultraviolet (UV)-inactivated (noninfectious) influenza virus in tissue culture. The "conditioned" medium from these experiments (which did not contain infectious virus) was found to inhibit the infection of fresh pieces of chick chorioallantoic membrane by (infectious) influenza virus in separate cultures (Figure 6.6). Their conclusion was that a soluble factor, which they called "interferon," was produced by cells as a result of virus infection and that this factor could prevent the infection of other cells. As a result of this provocative observation, IFN became the great hope for virology and was thought to be directly equivalent to the use of antibiotics to treat bacterial infections.

The true situation has turned out to be far more complex than was first thought. IFNs do have antiviral properties, but by and large their effects are exerted indirectly via their major function as cellular regulatory proteins. IFNs are immensely potent; less than 50 molecules per cell show evidence of antiviral activity. Hence, following Isaacs and Lindenmann's initial discovery, many fairly fruitless years were spent trying to purify minute amounts of naturally produced IFN.

This situation changed with the development of molecular biology and the cloning and expression of IFN genes, which has led to rapid advances in our understanding over the last 15 years. There are a number of different types of IFNs:

- IFN-α: There are at least 15 molecular species of IFN-α, all of which are closely related; some species differ by only one amino acid. They are synthesized predominantly by lymphocytes. The mature proteins contain 143 amino acids, with a minimum homology of 77% between the different types. All the genes encoding IFN-α are located on human chromosome 9, and gene duplication is thought to be responsible for this proliferation of genes.
- IFN-β: The single gene for IFN-β is also located on human chromosome 9. The mature protein contains 145 amino acids and, unlike IFN-α, is glycosylated, with approximately 30% homology to other IFNs. It is synthesized predominantly by fibroblasts.
- Other IFNs: The single gene for IFN-γ is located on human chromosome 12. The mature protein contains 146 amino acids, is glycosylated, and has very low sequence homology to other IFNs. It is synthesized predominantly by lymphocytes. Other IFNs, such as IFN-γ, -δ, -k, -τ, etc., play a variety of roles in cellular regulation but are not directly involved in controlling virus infection.

Because there are clear biological differences between the two main types of IFN, IFN-α and -β are known as type I IFN, and IFN-γ as type II IFN. Induction of IFN synthesis results from upregulation of transcription from the IFN gene promoters. There are three main mechanisms involved:

- **Virus infection:** This mechanism is thought to act by the inhibition of cellular protein synthesis that occurs during many virus infections, resulting in a reduction in the concentration of intracellular repressor proteins and hence in increased IFN gene transcription. In general, RNA viruses are potent inducers of IFN while DNA viruses are relatively poor inducers; however, there are exceptions to this rule (e.g., poxviruses are very potent inducers). The molecular events in the induction of IFN synthesis by virus infection are not clear. In some cases (e.g., influenza virus), UV-inactivated virus is a potent inducer; therefore, virus replication is not necessarily required. Induction by viruses might involve perturbation of the normal cellular environment and/or production of small amounts of double-stranded RNA.
- **Double-stranded (ds) RNA:** All naturally occurring double-stranded RNAs (e.g., reovirus genomes) are potent inducers of IFN, as are synthetic molecules (e.g., poly I:C); therefore, this process is

independent of nucleotide sequence. Single-stranded RNA and double-stranded DNA are not inducers. This mechanism of induction is thought to depend on the secondary structure of the RNA rather than any particular nucleotide sequence.

■ **Metabolic inhibitors:** Compounds that inhibit transcription (e.g., actinomycin D) or translation (e.g., cycloheximide) result in induction of IFN. Tumor promoters such as tetradecanoyl phorbol acetate or dimethyl sulfoxide are also inducers. Their mechanism of action remains unknown but they almost certainly act at the level of transcription.

The effects of IFNs are exerted via specific receptors that are ubiquitous on nearly all cell types (therefore, nearly all cells are potentially IFN responsive). There are distinct receptors for type I and type II IFN, each of which consists of two polypeptide chains. Binding of IFN to the type I receptor activates a specific cytoplasmic tyrosine kinase (Janus kinase, or Jak1), which phosphorylates another cellular protein, signal transducer and activator of transcription 2 (STAT2). This is transported to the nucleus and turns on transcriptional activation of IFN-responsive genes (including IFN, resulting in amplification of the original signal). Binding of IFN to the type II receptor activates a different cytoplasmic tyrosine kinase (Jak2), which phosphorylates the cellular protein STAT1, leading to transcriptional activation of a different set of genes.

The main action of IFNs is on cellular regulatory activities and is rather complex. IFN affects both cellular proliferation and immunomodulation. These effects result from the induction of transcription of a wide variety of cellular genes, including other cytokines. The net result is complex regulation of the ability of a cell to proliferate, differentiate, and communicate. This cell-regulatory activity itself has indirect effects on virus replication. Type I IFN is the major antiviral mechanism—other IFNs act as potent cellular regulators, which may have indirect antiviral effects in some circumstances.

The effect of IFNs on virus infections *in vivo* is extremely important. Animals experimentally infected with viruses and injected with anti-IFN antibodies experience much more severe infections than control animals infected with the same virus. This is because IFNs protect cells from damage and death. However, they do not appear to play a major role in the clearance of virus infections—the other parts of the immune response are necessary for this. IFN is a "firebreak" that inhibits virus replication in its earliest stages by several mechanisms. Two of these are understood in some detail, but a number of others (in some cases specific to certain viruses) are less well understood.

IFNs induce transcription of a cellular gene for the enzyme 2′,5′-oligo A synthetase (Figure 6.7). There are at least four molecular species of 2′,5′-oligo A,

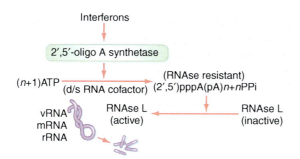

FIGURE 6.7 Induction of 2′,5′-oligo A synthetase by IFNs.
The modified nucleic acid 2′,5′-oligo A is involved in one of the major mechanism by which IFNs counteract virus infections.

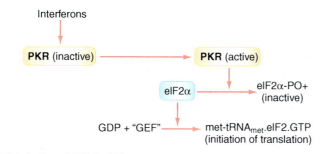

FIGURE 6.8 Induction of PKR by IFNs.
The protein kinase PKR is another major mechanism by which IFNs counteract virus infections.

induced by different forms of IFN. This compound activates an RNA-digesting enzyme, RNAse L, which digests virus genomic RNAs, virus and cellular mRNAs, and cellular ribosomal RNAs. The end result of this mechanism is a reduction in protein synthesis (due to the degradation of mRNAs and rRNAs)—therefore the cell is protected from virus damage. The second method relies on the activation of a 68-kDa protein called PKR (Figure 6.8). PKR phosphorylates a cellular factor, eIF2α, which is required by ribosomes for the initiation of translation. The net result of this mechanism is also the inhibition of protein synthesis and this reinforces the 2′,5′-oligo A mechanism. A third, well-established mechanism depends on the M_x gene, a single-copy gene located on human chromosome 21, the transcription of which is induced by type I IFN. The product of this gene inhibits the primary transcription of influenza virus but not of other viruses. Its method of action is unknown. In addition to these three mechanisms, there

Table 6.1 Therapeutic Uses of IFNs

Condition	Virus
Chronic active hepatitis	HBV, HCV
Condylomata accuminata (genital warts)	Papillomaviruses
Tumors	
Hairy cell leukemia	—
Kaposi's sarcoma (in AIDS patients)	Human herpesvirus 8 (HHV-8) (?)
Congenital Diseases	
Chronic granulomatous disease (IFN-γ reduces bacterial infections)	—

are many additional recorded effects of IFNs. They inhibit the penetration and uncoating of SV40 and some other viruses, possibly by altering the composition or structure of the cell membrane; they inhibit the primary transcription of many virus genomes (e.g., SV40, HSV) and also cell transformation by retroviruses. None of the molecular mechanisms by which these effects are mediated has been fully explained.

IFNs are a powerful weapon against virus infection, but they act as a blunderbuss rather than a "magic bullet." The severe side effects (fever, nausea, malaise) that result from the powerful cell-regulatory action of IFNs means that they will never be widely used for the treatment of trivial virus infections—they are not the cure for the common cold. However, as the cell-regulatory potential of IFNs is becoming better understood, they are finding increasing use as a treatment for certain cancers (e.g., the use of IFN-α in the treatment of hairy cell leukemia). Current therapeutic uses of IFNs are summarized in Table 6.1. The long-term prospects for their use as antiviral compounds are less certain, except for possibly in life-threatening infections where there is no alternative therapy (e.g., chronic viral hepatitis).

EVASION OF IMMUNE RESPONSES BY VIRUSES

In total, the many innate and adaptive components of the immune system present a powerful barrier to virus replication. Simply by virtue of their continued existence, it is obvious that viruses have, over millennia, evolved effective "counter-surveillance" mechanisms in this molecular arms race.

Inhibition of MHC-I-Restricted Antigen Presentation

As described above, CTLs can only respond to foreign antigens presented by MHC-I complexes on the target cell. A number of viruses interfere with MHC-I expression or function to disrupt this process and evade the CTL response. Such mechanisms include downregulation of MHC-I expression by adenoviruses and interference with the antigen processing required to form an MHC-I—antigen complex by herpesviruses.

Inhibition of MHC-II-Restricted Antigen Presentation

The MHC-II antigens are essential in the adaptive immune response in order to stimulate the development of antigen-responsive clones of effector cells. Again, herpesviruses and papillomaviruses interfere with the processing and surface expression of MHC-II—antigen complexes, inhibiting the CTL response.

Inhibition of NK Cell Lysis

The poxvirus *Molluscum contagiosum* encodes a homologue of MHC-I that is expressed on the surface of infected cells but is unable to bind an antigenic peptide, thus avoiding killing by NK cells that would be triggered by the absence of MHC-I on the cell surface. Similar proteins are made by other viruses, such as HHV-5 (CMV), and herpesviruses in general appear to have a number of sophisticated mechanisms to avoid NK cell killing.

Interference with Apoptosis

See Viruses and Apoptosis earlier in this chapter.

Inhibition of Cytokine Action

Cytokines are secreted polypeptides that coordinate important aspects of the immune response, including inflammation, cellular activation, proliferation, differentiation, and chemotaxis. Some viruses are able to inhibit the expression of certain chemokines directly. Alternatively, herpesviruses and poxviruses encode "viroceptors"—virus homologues of host cytokine receptors that compete with cellular receptors for cytokine binding but fail to give transmembrane signals. High-affinity binding molecules may also neutralize cytokines directly, and molecules known as "virokines" block cytokine receptors again without activating the intracellular signaling cascade.

IFNs are cytokines which act as an effective means of curbing the worst effects of virus infections. Part of their wide-ranging efficacy results from their generalized, nonspecific effects (e.g., the inhibition of protein synthesis in virus-infected cells). This lack of specificity means that it is very difficult for viruses to evolve strategies to counteract their effects; nevertheless, there are instances where this has happened. The anti-IFN effect of adenovirus VA RNAs has already been described in Chapter 5. Other mechanisms of virus resistance to IFNs include:

- EBV EBER RNAs are similar in structure and function to the adenovirus VA RNAs. The EBNA-2 protein also blocks IFN-induced signal transduction.
- Vaccinia virus (VV) is known to show resistance to the antiviral effects of IFNs. One of the early genes of this virus, K3L, encodes a protein that is homologous to eIF-2α, which inhibits the action of PKR. In addition, the E3L protein also binds dsRNA and inhibits PKR activation.
- Poliovirus infection activates a cellular inhibitor of PKR in virus-infected cells.
- Reovirus capsid protein σ3 is believed to sequester dsRNA and therefore prevent activation of PKR.
- Influenza virus NS1 protein suppresses IFN induction by blocking signaling through the Jak/STAT system.

Evasion of Humoral Immunity

Although direct humoral immunity is less significant than cell-mediated immunity, the antiviral action of ADCC and complement make this a worthwhile target to inhibit. The most frequent means of subverting the humoral response is by high-frequency genetic variation of the B-cell epitopes on antigens to which antibodies bind. This is only possible for viruses that are genetically variable (e.g., influenza virus and HIV). Herpesviruses use alternative strategies such as encoding viral Fc receptors to prevent Fc-dependent immune activation.

Evasion of the Complement Cascade

Poxviruses, herpesviruses, and some retroviruses encode mimics of normal regulators of complement activation proteins (e.g., secreted proteins that block C3 convertase assembly and accelerate its decay). Poxviruses can also inhibit C9 polymerization, preventing membrane permeabilization.

VIRUS–HOST INTERACTIONS

Viruses do not set out to kill their hosts. Virus pathogenesis is an abnormal situation of no value to the virus—the vast majority of virus infections are asymptomatic. However, for pathogenic viruses, a number of critical stages in

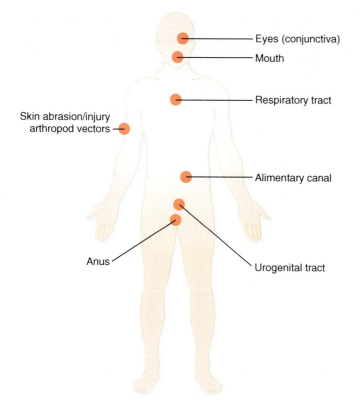

FIGURE 6.9 Sites of virus entry into the body.
The course a virus infection follows depends on the biology of the virus and the response to infection by the host, but is also influenced by the site at which the virus enters the body.

replication determine the nature of the disease they produce. For all viruses, pathogenic or nonpathogenic, the first factor that influences the course of infection is the mechanism and site of entry into the body (Figure 6.9):

- **The skin:** Mammalian skin is a highly effective barrier against viruses. The outer layer (epidermis) consists of dead cells and therefore does not support virus replication. Very few viruses infect directly by this route unless there is prior injury such as minor trauma or puncture of the barrier, such as insect or animal bites or subcutaneous injections. Some viruses that do use this route include HSV and papillomaviruses, although these viruses probably still require some form of disruption of the skin such as small abrasions or eczema.
- **Mucosal membranes:** The mucosal membranes of the eye and genitourinary (GU) tract are much more favorable routes of access for viruses to the tissues of the body. This is reflected by the number of

Table 6.2 Viruses that Infect via Mucosal Surfaces

Virus	Site of Infection
Adenoviruses	Conjunctiva
Picornaviruses—enterovirus 70	Conjunctiva
Papillomaviruses	GU tract
Herpesviruses	GU tract
Retroviruses—HIV, human T-cell leukemia virus (HTLV)	GU tract

Table 6.3 Viruses that Infect via the Alimentary Canal

Virus	Site of Infection
Herpesviruses	Mouth and oropharynx
Adenoviruses	Intestinal tract
Caliciviruses	Intestinal tract
Coronaviruses	Intestinal tract
Picornaviruses—enteroviruses	Intestinal tract
Reoviruses	Intestinal tract

viruses that can be sexually transmitted; virus infections of the eye are also quite common (Table 6.2).

- **Alimentary canal:** Viruses may infect the alimentary canal via the mouth, oropharynx, gut, or rectum, although viruses that infect the gut via the oral route must survive passage through the stomach, an extremely hostile environment with a very low pH and high concentrations of digestive enzymes. Nevertheless, the gut is a highly valued prize for viruses—the intestinal epithelium is constantly replicating and a good deal of lymphoid tissue is associated with the gut which provides many opportunities for virus replication. Moreover, the constant intake of food and fluids provides ample opportunity for viruses to infect these tissues (Table 6.3). To counteract this problem, the gut has many specific (e.g., secretory antibodies) and nonspecific (e.g., stomach acids and bile salts) defense mechanisms.
- **Respiratory tract:** The respiratory tract is probably the most frequent site of virus infection. As with the gut, it is constantly in contact with external virus particles which are taken in during respiration. As a result, the respiratory tract also has defenses aimed at virus infection—filtering of particulate matter in the sinuses and the presence of cells and antibodies of the immune system in the lower regions. Viruses that infect the respiratory tract usually come directly from the respiratory tract of others, as aerosol spread is very efficient: "coughs and sneezes spread diseases" (Table 6.4).

Table 6.4 Viruses that Infect via the Respiratory Tract

Virus	Localized Infection
Adenoviruses	Upper respiratory tract
Coronaviruses	Upper respiratory tract
Orthomyxoviruses	Upper respiratory tract
Picornaviruses—rhinoviruses	Upper respiratory tract
Paramyxoviruses—parainfluenza, respiratory syncytial virus	Lower respiratory tract

Virus	Systemic Infection
Herpesviruses	Varicella—Zoster
Paramyxoviruses	Measles, mumps
Poxviruses	Smallpox
Togaviruses	Rubella

The natural environment is a considerable barrier to virus infection. Most viruses are relatively sensitive to heat, drying, UV light (sunlight), etc., although a few types are quite resistant to these factors. This is particularly important for viruses that are spread via contaminated water or foodstuffs—not only must they be able to survive in the environment until they are ingested by another host, but, as most are spread by the fecal—oral route, they must also be able to pass through the stomach to infect the gut before being shed in the feces. One way of overcoming environmental stress is to take advantage of a secondary vector for transmission between the primary hosts (Figure 6.10). As with plant viruses, the virus may or may not replicate while in the vector. Viruses without a secondary vector must rely on continued host-to-host transmission and have evolved various strategies to do this (Table 6.5):

- **Horizontal transmission:** The direct host-to-host transmission of viruses. This strategy relies on a high rate of infection to maintain the virus population.
- **Vertical transmission:** The transmission of the virus from one generation of hosts to the next. This may occur by infection of the fetus before, during, or shortly after birth (e.g., during breastfeeding). More rarely, it may involve direct transfer of the virus via the germ line itself (e.g., retroviruses). In contrast to horizontal transmission, this strategy relies on long-term persistence of the virus in the host rather than rapid propagation and dissemination of the virus.

Having gained entry to a potential host, the virus must initiate an infection by entering a susceptible cell (primary replication). This initial interaction frequently determines whether the infection will remain localized at the site

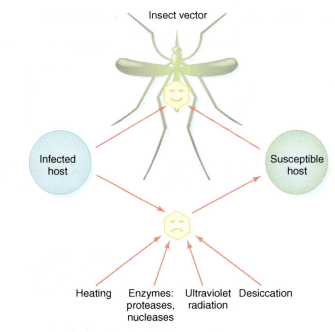

FIGURE 6.10 Transmission of viruses through the environment.
Some viruses have adopted the use of vectors such as insects or other arthropods to avoid environmental stresses when outside their host organism.

Table 6.5 Virus Transmission Patterns	
Pattern	**Example**
Horizontal Transmission	
Human–human (aerosol)	Influenza
Human–human (fecal–oral)	Rotaviruses
Animal–human (direct)	Rabies
Animal–human (vector)	Bunyaviruses
Vertical Transmission	
Placental–fetal	Rubella
Mother–child (birth)	HSV, HIV
Mother–child (breastfeeding)	HIV, HTLV
Germ line	In mice, retroviruses; in humans (?)

Table 6.6 Examples of Localized and Systemic Virus Infections

Virus	Primary Replication	Secondary Replication
Localized Infections		
Papillomaviruses	Dermis	—
Rhinoviruses	Upper respiratory tract	—
Rotaviruses	Intestinal epithelium	—
Systemic Infections		
Enteroviruses	Intestinal epithelium	Lymphoid tissues, CNS
Herpesviruses	Oropharynx or GU tract	Lymphoid cells, CNS

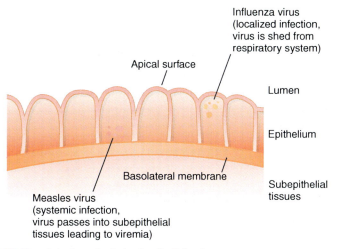

FIGURE 6.11 Virus infection of polarized epithelial cells.
Some viruses which infect epithelial cells are released from the apical surface (e.g., influenza virus) while others are released from the basolateral surface of the cells (e.g., rhabdoviruses). This affects the way in which the virus spreads through the body and the subsequent course of the infection.

of entry or spread to become a systemic infection (Table 6.6). In some cases, virus spread is controlled by infection of polarized epithelial cells and the preferential release of virus from either the apical (e.g., influenza virus—a localized infection in the upper respiratory tract) or basolateral (e.g., rhabdoviruses—a systemic infection) surface of the cells (Figure 6.11). Following primary replication at the site of infection, the next stage may be spread throughout the host. In addition to direct cell–cell contact, there are two main mechanisms for spread throughout the host:

- **Via the bloodstream:** Viruses may get into the bloodstream by direct inoculation—for example, by arthropod vectors, blood transfusion, or intravenous drug abuse (sharing of nonsterilized needles). The virus may travel free in the plasma (e.g., togaviruses, enteroviruses) or in association with red cells (orbiviruses), platelets (HSV), lymphocytes (EBV, CMV), or monocytes (lentiviruses). Primary viremia usually precedes and is necessary for the spread of virus to other parts of the body via the bloodstream and is followed by a more generalized, higher titer secondary viremia as the virus reaches the other target tissues or replicates directly in blood cells.
- **Via the nervous system:** As above, spread of virus to the nervous system is usually preceded by primary viremia. In some cases, spread occurs directly by contact with neurones at the primary site of infection; in other cases, it occurs via the bloodstream. Once in peripheral nerves, the virus can spread to the central nervous system (CNS) by axonal transport along neurones. The classic example of this is HSV (see "Latent Infection," below). Viruses can cross synaptic junctions as these frequently contain virus receptors, allowing the virus to jump from one cell to another.

The spread of the virus to various parts of the body is controlled to a large extent by its cell or tissue tropism. Tissue tropism is controlled partly by the route of infection but largely by the interaction of a virus-attachment protein (VAP) with a specific receptor molecule on the surface of a cell (as discussed in Chapter 4) and has considerable effect on pathogenesis.

At this stage, following significant virus replication and the production of virus antigens, the host immune response comes into play. This has already been discussed earlier and obviously has a major impact on the outcome of an infection. To a large extent, the efficiency of the immune response determines the amount of secondary replication that occurs and, hence, the spread to other parts of the body. If a virus can be prevented from reaching tissues where secondary replication can occur, generally no disease results, although there are some exceptions to this. The immune response also plays a large part in determining the amount of cell and tissue damage that occurs as a result of virus replication. As described above, the production of IFNs is a major factor in preventing virus-induced tissue damage.

The immune system is not the only factor that controls cell death, the amount of which varies considerably for different viruses. Viruses may replicate widely throughout the body without any disease symptoms if they do not cause significant cell damage or death. Retroviruses do not generally cause cell death, being released from the cell by budding rather than by cell lysis, and cause persistent infections, even being passed vertically to the offspring if they infect the germ line. All vertebrate genomes, including humans, are littered with retrovirus genomes that have been with us for millions of years (Chapter 3). At present, these ancient virus genomes are not known to cause any disease in humans, although there are examples of tumors caused by them in rodents. Conversely, picornaviruses cause lysis and death of the cells in which they replicate, leading to fever and increased mucus secretion, in the case of rhinoviruses, and paralysis or death (usually due to respiratory failure due to damage to the CNS resulting, in part, from virus replication in these cells) in the case of poliovirus.

The eventual outcome of any virus infection depends on a balance between two processes. Clearance is mediated by the immune system (as discussed previously); however, the virus is a moving target that responds rapidly to pressure from the immune system by altering its antigenic composition (whenever possible). The classic example of this phenomenon is influenza virus, which displays two genetic mechanisms that allow the virus to alter its antigenic constitution:

- **Antigenic drift:** This involves the gradual accumulation of minor mutations (e.g., nucleotide substitutions) in the virus genome which result in subtly altered coding potential and therefore altered antigenicity, leading to decreased recognition by the immune system. This process occurs in all viruses all the time but at greatly different rates; for example, it is much more frequent in RNA viruses than in DNA viruses. In response, the immune system constantly adapts by recognition of and response to novel antigenic structures—but it is always one step behind. In most cases, however, the immune system is eventually able to overwhelm the virus, resulting in clearance.
- **Antigenic shift:** In this process, a sudden and dramatic change in the antigenicity of a virus occurs owing to reassortment of the segmented virus genome with another genome of a different antigenic type (see Chapter 3). This results initially in the failure of the immune system to recognize a new antigenic type, giving the virus the upper hand (Figure 6.12).

The occurrence of past antigenic shifts in influenza virus populations is recorded by pandemics (worldwide epidemics; Figure 6.13). These events are marked by the sudden introduction of a new antigenic type of hemagglutinin and/or neuraminidase into the circulating virus, overcoming previous immunity in the human population. Previous hemagglutinin/neuraminidase types

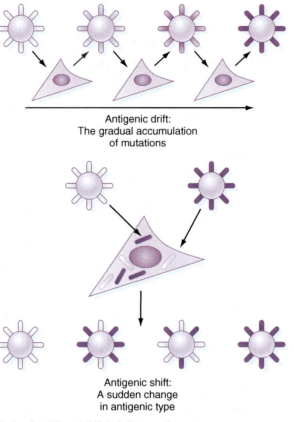

Antigenic drift:
The gradual accumulation
of mutations

Antigenic shift:
A sudden change
in antigenic type

FIGURE 6.12 **Antigenic shift and drift in influenza virus.**
Variation in the antigenicity of influenza viruses occurs through two mechanisms, gradual antigenic drift
and sudden antigenic shifts.

become resurgent when a sufficiently high proportion of the people who
have "immunological memory" of that type have died, thus overcoming the
effect of "herd immunity."

The other side of the relationship that determines the eventual outcome of a
virus infection is the ability of the virus to persist in the host. Long-term
persistence of viruses results from two main mechanisms. The first is the
regulation of lytic potential. The strategy followed here is to achieve the continued survival of a critical number of virus-infected cells (i.e., sufficient to
continue the infection without killing the host organism). For viruses that do
not usually kill the cells in which they replicate, this is not usually a problem; hence, these viruses tend naturally to cause persistent infections (e.g.,
retroviruses). For viruses that undergo lytic infection (e.g., herpesviruses), it

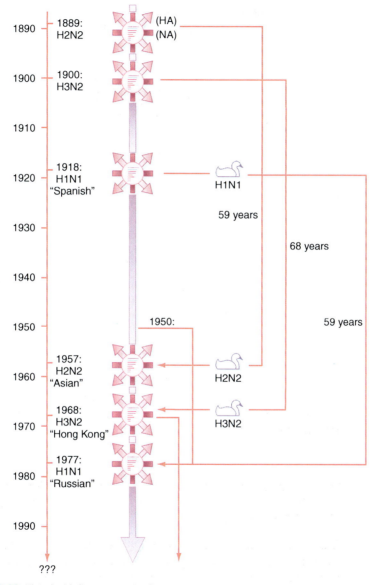

FIGURE 6.13 Historical influenza pandemics.
This chart shows the history of influenza pandemics throughout the twentieth century. The first pandemic of the twenty-first century occurred in 2009 and was caused by an H1N1 type virus, although this was not as damaging as earlier pandemics.

is necessary to develop mechanisms that restrict virus gene expression and, consequently, cell damage. The second aspect of persistence is the evasion of immune surveillance, discussed above.

THE COURSE OF VIRUS INFECTIONS

Patterns of virus infection can be divided into a number of different types.

Abortive Infection

Abortive infection occurs when a virus infects a cell (or host) but cannot complete the full replication cycle, so this is a nonproductive infection. The outcome of such infections is not necessarily insignificant, for example, SV40 infection of nonpermissive rodent cells sometimes results in transformation of the cells (see Chapter 7).

Acute Infection

This pattern is familiar for many common virus infections (e.g., "colds"). In these relatively brief infections, the virus is usually eliminated completely by the immune system. Typically, in acute infections, much virus replication occurs before the onset of any symptoms (e.g., fever), which are the result not only of virus replication but also of the activation of the immune system; therefore, acute infections present a serious problem for the epidemiologist and are the pattern most frequently associated with epidemics (e.g., influenza, measles).

Chronic Infection

These are the converse of acute infections (i.e., prolonged and stubborn). To cause this type of infection, the virus must persist in the host for a significant period. To the clinician, there is no clear distinction among chronic, persistent, and latent infections, and the terms are often used interchangeably. They are listed separately here because, to virologists, there are significant differences in the events that occur during these infections.

Persistent Infection

These infections result from a delicate balance between the virus and the host organism, in which ongoing virus replication occurs but the virus adjusts its replication and pathogenicity to avoid killing the host. In chronic infections, the virus is usually eventually cleared by the host (unless the infection proves fatal), but in persistent infections the virus may continue to be present and to replicate in the host for its entire lifetime.

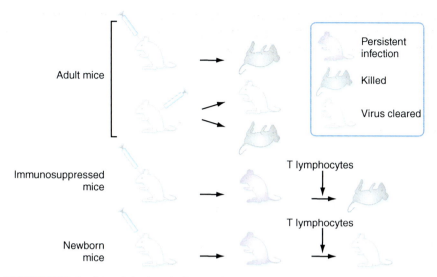

Adult mice

Persistent infection

Killed

Virus cleared

Immunosuppressed mice

T lymphocytes

Newborn mice

T lymphocytes

FIGURE 6.14 Persistent infection of mice by LCMV.
LCMV is an arenavirus where the course of infection depends in part on the immune response of the host animal to the virus.

The best-studied example of such a system is lymphocytic choriomeningitis virus (LCMV; an arenavirus) infection in mice (Figure 6.14). Mice can be experimentally infected with this virus either at a peripheral site (e.g., a footpad or the tail) or by direct inoculation into the brain. Adult mice infected in the latter way are killed by the virus, but among those infected by a peripheral route there are two possible outcomes to the infection: some mice die but others survive, having cleared the virus from the body completely. It is not clear what factors determine the survival or death of LCMV-infected mice, but other evidence shows that the outcome is related to the immune response to the virus. In immunosuppressed adult mice infected via the CNS route, a persistent infection is established in which the virus is not cleared (due to the nonfunctional immune system), but, remarkably, these mice are not killed by the virus. If, however, syngeneic LCMV-specific T-lymphocytes (i.e., of the same MHC type) are injected into these persistently infected mice, the animals develop the full pathogenic symptoms of LCMV infection and die. When newborn mice, whose immune systems are immature, are infected via the CNS route, they also develop a persistent infection, but, in this case, if they are subsequently injected with syngeneic LCMV-specific T-lymphocytes, they clear the virus and survive the infection. The mechanisms that control these events are not completely understood, but evidently there is a delicate balance between the virus and the host animal and the immune response to the virus is partly responsible for the pathology of the disease and the death of the animals.

Not infrequently, persistent infections may result from the production of defective-interfering (D.I.) particles (see Chapter 3). Such particles contain a partial deletion of the virus genome and are replication defective, but they are maintained and may even tend to accumulate during infections because they can replicate in the presence of replication-competent helper virus. The production of D.I. particles is a common consequence of virus infection of animals, particularly by RNA viruses, but also occurs with DNA viruses and plant viruses and can be mimicked *in vitro* by continuous high-titer passage of virus. Although not able to replicate themselves independently, D.I. particles are not necessarily genetically inert and may alter the course of an infection by recombination with the genome of a replication-competent virus. The presence of D.I. particles can profoundly influence the course and the outcome of a virus infection. In some cases, they appear to moderate pathogenesis, whereas in others they potentiate it, making the symptoms of the disease much more severe. Moreover, as D.I. particles effectively cause restricted gene expression (because they are genetically deleted), they may also result in a persistent infection by a virus that normally causes an acute infection and is rapidly cleared from the body.

Latent Infection

In a latent state, the virus is able to downregulate its gene expression and enter an inactive state with strictly limited gene expression and without ongoing virus replication. Latent virus infections typically persist for the entire life of the host. An example of such an infection in humans is HSV. Infection of sensory nerves serving the mucosa results in localized primary replication. Subsequently, the virus travels via axon transport mechanisms further into the nervous system. There, it hides in dorsal root ganglia, such as the trigeminal ganglion, establishing a truly latent infection. The nervous system is an immunologically privileged site and is not patrolled by the immune system in the same way as the rest of the body, but the major factor in latency is the ability of the virus to restrict its gene expression. This eliminates the possibility of recognition of infected cells by the immune system. Restricted gene expression is achieved by tight regulation of α-gene expression, which is an essential control point in herpesvirus replication (Chapter 5). In the latent state, HSV makes an 8.3-kb RNA transcript called the latent RNA or latency-associated transcript (LAT). The LAT is broken down into even smaller strands called microRNAs (miRNAs), and these block the production of proteins which reactivate the virus. Drugs which block production of these miRNAs could in theory "wake up" all the dormant viruses, making them vulnerable to the immune system and to antiviral therapy, and this raises the eventual possibility of a cure for herpes infections. Expression of the LAT promotes neuronal survival after HSV infection by inhibiting apoptosis. This anti-apoptosis function could promote reactivation by:

- Providing more latently infected neurons for future reactivations,
- Protecting neurons in which reactivation occurs,
- Protecting previously uninfected neurons during a reactivation.

When reactivated by some provocative stimulus, HSV travels down the sensory nerves to cause peripheral manifestations such as cold sores or genital ulcers. It is not altogether clear what constitutes a provocative stimulus, but there are many possible alternatives, including psychological and physical factors. Periodic reactivation establishes the pattern of infection, with sporadic, sometimes very painful reappearance of disease symptoms for the rest of the host's life. Even worse than this, immunosuppression later in life can cause the latent infection to flare up (which indicates that the immune system normally has a role in helping to suppress these latent infections), resulting in a very severe, systemic, and sometimes life-threatening infection.

In a manner somewhat similar to herpesviruses, infection by retroviruses may result in a latent infection. Integration of the provirus into the host genome certainly results in the persistence of the virus for the lifetime of the host organism and may lead to an episodic pattern of disease. In some ways, acquired immunodeficiency syndrome (AIDS), which results from HIV infection, shows aspects of this pattern of infection. The pathogenesis of AIDS is discussed in detail in Chapter 7.

PREVENTION AND THERAPY OF VIRUS INFECTION

There are two aspects of the response to the threat of virus diseases: first, prevention of infection, and second, treatment of the disease. The former strategy relies on two approaches: public and personal hygiene, which perhaps plays the major role in preventing virus infection (e.g., provision of clean drinking water and disposal of sewage; good medical practice such as the sterilization of surgical instruments) and vaccination, which makes use of the immune system to combat virus infections. Most of the damage to cells during virus infections occurs very early, often before the clinical symptoms of disease appear. This makes the treatment of virus infection very difficult; therefore, in addition to being less expensive, prevention of virus infection is undoubtedly better than cure.

To design effective vaccines, it is important to understand both the immune response to virus infection and the stages of virus replication that are appropriate targets for immune intervention. To be effective, vaccines must stimulate as many of the body's defense mechanisms as possible. In practice, this usually means trying to mimic the disease without causing pathogenesis—for example, the use of live attenuated viruses as vaccines such as nasally administered influenza vaccines and orally administered poliovirus vaccines. To be effective, it is not necessary to get 100% uptake of vaccine. "Herd immunity" results from the break in transmission of a virus that occurs when a sufficiently high proportion of a population has been vaccinated. This strategy is most effective where there is no alternative host for the virus (e.g., measles) and in practice is the situation that usually occurs as it is impossible to achieve 100% coverage with any vaccine. However, this is a risky business; if protection of the population falls below a critical level, epidemics can easily occur.

Synthetic vaccines are short, chemically synthesized peptides. The major disadvantage with these molecules is that they are not usually very effective immunogens and are very costly to produce. However, because they can be made to order for any desired sequence, they have great theoretical potential, but none are currently in clinical use.

Recombinant vaccines are produced by genetic engineering. Such vaccines have been already produced and are better than synthetic vaccines because they tend to give rise to a more effective immune response. Some practical success has already been achieved with this type of vaccine. For example, vaccination against hepatitis B virus (HBV) used to rely on the use of Australian antigen (HBsAg) obtained from the serum of chronic HBV carriers. This was a very risky practice indeed (because HBV carriers are often also infected with HIV). A completely safe recombinant HBV vaccine produced in yeast is now used.

DNA vaccines are the newest type of vaccine and consist of only a DNA molecule encoding the antigen(s) of interest and, possibly, costimulatory molecules such as cytokines. The concept behind these vaccines is that the DNA component will be expressed *in vivo*, creating small amounts of antigenic protein that serve to prime the immune response so that a protective response can be rapidly generated when the real antigen is encountered. In theory, these vaccines could be manufactured quickly and should efficiently induce both humoral and cell-mediated immunity. Initial clinical studies have indicated that there is still some way to go until this experimental technology becomes a practical proposition.

Subunit vaccines consist of only some components of the virus, sufficient to induce a protective immune response but not enough to allow any danger of infection. In general terms, they are completely safe, except for very rare cases in which adverse immune reactions may occur. Unfortunately, they also tend to be the least effective and most expensive type of vaccine. The major technical problems associated with subunit vaccines are their relatively poor antigenicity and the need for new delivery systems, such as improved carriers and adjuvants.

Virus vectors are recombinant virus **genomes** genetically manipulated to express protective antigens from (unrelated) pathogenic viruses. The idea here is to utilize the genome of a well-understood, attenuated virus to express and present antigens to the immune system. Many different viruses offer possibilities for this type of approach. One of the most highly developed systems so far is based on the VV genome. This virus has been used to vaccinate millions of people worldwide in the campaign to eradicate smallpox and is generally a safe and effective vehicle for antigen delivery. Such vaccines are difficult to produce. No human example is clearly successful yet, although many different trials are currently underway, but VV–rabies recombinants have been used to eradicate rabies in European fox populations. VV-based vaccines have advantages and disadvantages for use in humans—a high percentage of the human population has already been vaccinated during the smallpox eradication campaign, and this lifelong protection may result in poor response to recombinant vaccines. Although generally safe, VV is dangerous in immunocompromised hosts, thus it cannot be used in HIV-infected individuals. A possible solution to these problems may be to use avipoxvirus vectors (e.g., fowlpox or canarypox) as "suicide vectors" that can only establish **abortive infections** of mammalian cells and offer the following advantages:

- Expression of high levels of foreign proteins,
- No danger of pathogenesis (abortive infection),
- No natural immunity in humans (avian virus).

Inactivated vaccines are produced by exposing the virus to a denaturing agent under precisely controlled conditions. The objective is to cause loss of virus infectivity without loss of antigenicity. Obviously, this involves a delicate balance. However, inactivated vaccines have certain advantages, such as generally being effective immunogens (if properly inactivated), being relatively stable, and carrying little or no risk of vaccine-associated virus infection (if properly inactivated, but accidents can and do occur). The disadvantage of these vaccines is that it is not possible to produce inactivated vaccines for all viruses, as denaturation of virus proteins may lead to loss of antigenicity (e.g., measles virus). Although relatively effective, "killed" vaccines are sometimes not as effective at preventing infection as "live" virus vaccines, often because they fail to stimulate protective mucosal and cell-mediated immunity to the same extent. A more recent concern is that these vaccines contain virus nucleic acids, which may themselves be a source of infection, either of their own accord (e.g., (+)sense RNA virus **genomes**) or after **recombination** with other viruses.

Virus vaccines do not have to be based on **virion** structural proteins. The effectiveness of attenuated vaccines relies on the fact that a complete spectrum of virus proteins, including nonstructural proteins, is expressed and gives rise to cell-mediated immune responses. Live attenuated virus vaccines are viruses with reduced pathogenicity used to stimulate an immune response without causing disease. The vaccine strain may be a naturally occurring virus (e.g., the use of cowpox virus by Edward Jenner to vaccinate against smallpox) or artificially attenuated *in vitro* (e.g., the oral poliomyelitis vaccines produced by Albert Sabin). The advantage of attenuated vaccines is that they are good immunogens and induce long-lived, appropriate immunity. Set against this advantage are their many disadvantages. They are often biochemically and genetically unstable and may either lose infectivity (becoming worthless) or revert to virulence unexpectedly. Despite intensive study, it is not possible to produce an attenuated vaccine to order, and there appears to be no general mechanism by which different viruses can be reliably and safely attenuated. Contamination of the vaccine stock with other, possibly pathogenic viruses is also possible—this was the way in which SV40 was first discovered in oral poliovirus vaccine in 1960. Inappropriate use of live virus vaccines, for example, in immunocompromised hosts or during pregnancy may lead to vaccine-associated disease, whereas the same vaccine given to a healthy individual may be perfectly safe.

Despite these difficulties, vaccination against virus infection has been one of the great triumphs of medicine during the twentieth century. Most of the success stories result from the use of live attenuated vaccines—for example, the use of VV against smallpox. On May 8, 1980, the World Health Organization (WHO) officially declared smallpox to be completely eradicated, the first virus disease to be eliminated from the world. The WHO aims to eradicate a number of other virus diseases such as poliomyelitis and measles, but targets for completion of these programs have undergone much slippage due to the formidable difficulties involved in a worldwide undertaking of this nature.

Although prevention of infection by prophylactic vaccination is much the preferred option, postexposure therapeutic vaccines can be of great value in modifying the course of some virus infections. Examples of this include rabies virus, where the course of infection may be very long and there is time for postexposure vaccination to generate an effective immune response and prevent the virus from carrying out the secondary replication in the CNS that is responsible for the pathogenesis of rabies. Other potential examples can be found in virus-associated tumors, such as HPV-induced cervical carcinoma.

Most existing virus vaccines are directed against viruses which are relatively antigenically invariant, for example, measles, mumps, and rubella viruses, where this is only one unchanging serotype of the virus. Viruses whose antigenicity alters continuously are a major problem in terms of vaccine production, and the classic example of this is influenza virus (see earlier). In response to this problem, new technologies such as reverse genetics could be used to improve and to shorten the lengthy process of preparing vaccines. RNA virus genomes can be easily manipulated as DNA clones to contain nucleotide sequences which match currently circulating strains of the virus. Infectious virus particles are rescued from the DNA clones by introducing these into cells. Seed viruses for distribution to vaccine manufacturers can be produced in as little as 1–2 weeks, a much shorter time than the months this process takes in conventional vaccine manufacture. Using the same technology, universal influenza vaccines containing crucial virus antigens expressed as fusion proteins with other antigenic molecules could feasibly be produced, making the requirement for constant production of new influenza vaccines obsolete. Although this has not yet been achieved, advances toward these goals are being made. The explosion of molecular techniques described in earlier chapters is now being used to inform vaccine design (as well as the design of antiviral drugs) rather than simply relying on trial-and-error approaches. However, developing safe and effective vaccines remains one of the greatest challenges facing virology.

RNA INTERFERENCE

RNA interference (RNAi) is a posttranscriptional gene silencing process that occurs in organisms from yeast to humans. In mammals, small RNAs include small interfering RNAs (siRNAs) and miRNAs. siRNAs, with perfect base complementarity to their targets, activate RNAi-mediated cleavage of the target mRNAs, while miRNAs generally induce RNA decay and/or translation inhibition of target genes (Figure 6.15). Mammals, including humans, encode hundreds or thousands of miRNAs. Some viruses with eukaryotic hosts also encode miRNAs. Herpesviruses in particular encode multiple miRNAs; most other nuclear DNA viruses encode one or two miRNAs. RNA viruses and cytoplasmic DNA viruses appear to lack any miRNAs. Virus miRNAs may serve two major functions. Several have been shown to inhibit the expression of cellular factors that play a role in cellular innate or adaptive antiviral immune responses, so reducing the effectiveness of the immune response. Alternatively, virus miRNAs may downregulate the expression of virus proteins, including key immediate-early or early regulatory proteins. In HSV, miRNAs are expressed at high levels during latency, but not during productive replication, so their action is thought to stabilize latency.

Recently there have been controversial claims that miRNA can exert antiviral activity in mice, at least in some circumstances. Antiviral siRNA activity is only seen in stem cells and in newborn mice and many scientists think siRNA is not a major part of the innate immune system in adult animals. There is evidence that siRNA may be turned on in responses to virus

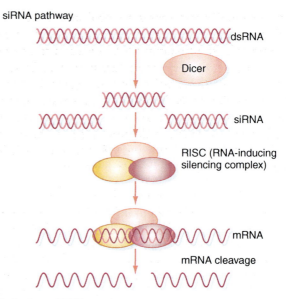

FIGURE 6.15 Mechanism of RNAi.

siRNAs have base complementarity to their target RNA molecules. The resulting double-stranded RNAs are processed by various enzymes, notably Dicer, to produce a complex (RISC) which carries out cleavage of the target mRNAs.

infection, but rather than acting directly against the virus, it may be used to regulate the IFN response. That still leaves the fact that in mammals miRNA is a powerful regulator of gene expression, including virus genes. Many viruses use miRNA to control their own gene expression and that of their host cells. On infection of a host cell, viruses encounter a range of miRNA species, many of which have been shown to restrict virus gene expression. Thus they have had to evolve a range of mechanisms to evade miRNA restriction is the same way that they have evolved other mechanisms to mitigate the impact of innate immunity. These include:

- Blocking miRNA function
- Avoiding 3′UTR targets complementary to cellular miRNAs
- Evolving very short 3′UTRs
- Evolving structured 3′UTRs

RNAi expression can be induced by dsRNA, and this approach has been used to investigate gene function in a variety of organisms including plants and insects. However, this method cannot be applied to mammalian cells as dsRNAs longer than 30 nucleotides induces the IFN response (see earlier), which results in the degradation of mRNAs and causes a global inhibition of translation. To circumvent this problem, chemically synthesized siRNAs or plasmid-vectors manipulated to produce short hairpin RNA molecules can be used to investigate gene function in mammals. In the future it may be feasible to treat virus diseases by shutting off gene expression by directing the degradation of specific mRNAs, and many clinical trials are currently underway. Although RNA interference has been used widely in cultured cells to inhibit virus replication and to probe biological pathways, considerable problems must be overcome before it becomes a useful therapy, including the development of suitable delivery and targeting systems and solving the issue of stability *in vivo*.

The natural world is a soup of bacteriophages. So how do bacteria survive against this constant onslaught? With their own form of adaptive immunity. CRISPRs (Clustered Regularly Interspaced Short Palindromic Repeats) are short, direct repeats of DNA base sequences. Each CRISPR contains a series of bases followed by the same series in reverse (a palindrome) and then by 30 or so base pairs known as "spacer" DNA. The spacers are short segments of virus or plasmid DNA. CRISPRs are found in the genomes of approximately 50% of bacteria and 90% of archaea. CRISPR loci are typically located on the bacterial chromosome, although some are found on plasmids. Bacteria may contain more than one CRISPR locus—up to 18 in some cases. CRISPRs function as a sort of prokaryotic immune system, conferring resistance to exogenous genetic elements such as plasmids and bacteriophages. Intriguingly, the CRISPR system provides a form of acquired immunity, allowing the cell to remember and respond to sequences it has encountered before.

How do CRISPRs work? CRISPRs are often adjacent to *cas* (CRISPR-associated) genes. The *cas* genes encode a large and heterogeneous family of proteins including nucleases, helicases, polymerases, and polynucleotide-binding proteins, forming the CRISPR/Cas system. (Note: *cas* = genes, Cas = the proteins encoded by these genes.) The interesting bits are the unique spacer elements (derived from exogenous sequences such as viruses and plasmids) rather than the repeats themselves. The spacer elements originate from exogenous DNA the bacterium (or its ancestors) has previous encountered—they are typically pieces of phage or plasmid DNA. This allows the cell to recognize these sequences via base homology if they enter the cell again, for example, if the bacterium is infected with a bacteriophage whose genome contains this sequence:

1. cas-encoded nucleases cleave invading DNA into short pieces.
2. Other cas proteins allow a fragment of the foreign DNA to be incorporated as a novel repeat-spacer unit at the leader end of the CRISPR site.

3. The CRISPR array is then transcribed to form a pre-CRISPR RNA (crRNA) transcript.
4. The pre-crRNA is cleaved within the repeat sequence by Cas proteins to generate small CRISPR RNAs, crRNAs.
5. The crRNAs to work in a similar way to RNAi in eukaryotic cells, although there are important differences in the machinery by which this happens.

CRISPRs are an important way in which bacteria are able to survive constant attack by bacteriophages in the environment, but phages have been around for a very long time too, so they must have found ways of counteracting the CRISPR system. Eukaryotic viruses may express inhibitors such as dsRNA-binding proteins that interfere with the RNA silencing machinery, whereas bacteriophages acquire mutations or recombine the sequence corresponding to the CRISPR spacer to avoid recognition in an analogous way to how viruses of eukaryotes acquire mutations in B-cell and T-cell epitopes in proteins to evade the mammalian immune system.

So who (apart from bacteria) cares about CRISPRs? Altering the spacer via genetic manipulation can provide novel phage resistance, whereas spacer deletion results in loss of phage resistance. Although CRISPRs originate in bacteria, they also work in eukaryotic cells if introduced by genetic engineering. This provides a convenient way of targeting genes in cells, including human cells. Recent work suggests that CRISPRs might also be involved in control of bacterial gene expression as well as in immunity. We will undoubtedly see much more widespread use of CRISPRs in biotechnology over the next few years.

VIRUSES AS THERAPEUTICS

Phage therapy, the use of bacteriophages to treat or prevent disease, stretches back a century to the earliest days of the discovery of phages. Long before the discovery of antibiotics, the thought that viruses which lyse bacteria could be used to treat diseases was highly attractive. Yet this idea has never become a widespread practical reality. Devotees of phage therapy defend their cherished belief with almost religious fervor, but there are serious obstacles to be overcome, such as the narrow host range of most phages (a few strains of bacteria, not even an entire species) and the speed at which bacteria develop resistance to infection. As the spectrum of clinically useful antibiotics dwindles, phage therapy increases in attractiveness, but is unlikely ever to replace the antibiotic golden era of disease treatment we are now leaving behind.

Another aspect of "virotherapy" is the growing interest in oncolytic viruses—viruses engineered to kill only cancer cells. The usefulness of many different types of virus has been investigated, including adenoviruses, herpesviruses, reoviruses, and poxviruses. Although safety is a concern even in patients with terminal illnesses, this is one area of medical research where optimism is considerable. Many clinical trials are underway at it seems certain that this approach to cancer treatment will eventually become more common, possibly as an adjunct to other forms of therapy such as surgery, drugs, and radiotherapy.

Viruses have also developed as gene delivery systems for the treatment of inherited and acquired diseases. Gene therapy offers:

- Delivery of large biomolecules to cells,
- The possibility of targeting delivery to a specific cell type,
- High potency of action due to replication of the vector,
- Potential to treat certain diseases (such as head and neck cancers and brain tumors) that respond poorly to other therapies or may be inoperable.

The very first retroviral and adenoviral vectors were characterized in the early 1980s. The first human trial to treat children with immunodeficiency resulting from a lack of the enzyme adenosine deaminase began in 1990 and showed encouraging although not completely successful results. Like most of the initial attempts, this trial used recombinant retrovirus genomes as vectors. In 1995, the first successful gene therapy for motor neurons and skin cells was reported, while the first phase three (widespread) gene therapy trial was begun in 1997. In 1999, the first successful treatment of a patient with severe combined immunodeficiency disease (SCID) was reported, but, sadly, the first death due to a virus vector also occurred, and in 2002 the occurrence of leukemias due to oncogenic insertion of a retroviral vector was seen in some SCID patients undergoing treatment. Several different viruses are being tested as potential vectors (Table 6.7). After initial optimism, gene therapy involving virus vectors has fallen from favor, and nonvirus methods of gene delivery including liposome/DNA complexes, peptide/DNA complexes, and direct injection of recombinant DNA are also under active investigation. It is important to note that such experiments are aimed at augmenting defective cellular genes in the somatic cells of patients to alleviate the symptoms of the disease and not at manipulating the human germ line, which is a different issue.

Table 6.7 Virus Vectors in Gene Therapy

Virus	Advantages	Possible Disadvantages
Adenoviruses	Relatively easily manipulated *in vitro* (cf. retroviruses); genes coupled to the major late promoter are efficiently expressed in large amounts.	Possible pathogenesis associated with partly attenuated vectors (especially in the lungs); immune response makes multiple doses ineffective if gene must be administered repeatedly (virus does not integrate).
Parvoviruses (AAV)	Integrate into cellular DNA at high frequency to establish a stable latent state; not associated with any known disease; vectors can be constructed that will not express any viral gene products.	Only ~5 kb of DNA can be packaged into the parvovirus capsid, and some virus sequences must be retained for packaging; integration into host-cell DNA may potentially have damaging consequences.
Herpesviruses	Relatively easy to manipulate *in vitro*; grows to high titers; long-term persistence in neuronal cells without integration.	(Long-term) pathogenic consequences?
Retroviruses	Integrate into cell genome, giving long-lasting (lifelong?) expression of recombinant gene.	Difficult to grow to high titer and purify for direct administration (patient cells must be cultured *in vitro*); cannot infect nondividing cells—most somatic cells (except lentiviruses?); insertional mutagenesis/activation of cellular oncogenes.
Poxviruses	Can express high levels of foreign proteins. Avipoxvirus vectors (e.g., fowlpox or canarypox) are "suicide vectors" that undergo abortive replication in mammalian cells so there is no danger of pathogenesis and no natural immunity in humans.	A high proportion of the human population has already been vaccinated—lifelong protection may result in poor response to recombinant vaccines (?). Dangerous in immunocompromised hosts.

CHEMOTHERAPY OF VIRUS INFECTIONS

BOX 6.3 THE DRUGS DON'T WORK

Pharmaceutical companies have a love–hate relationship with vaccines. Mostly hate. They are expensive and difficult to produce and save millions of lives, but if one child is harmed by an alleged bad reaction to a vaccination, the company suffers terrible publicity. Antiviral drugs however, now that's a different story. After suitable clinical trials antivirals are very safe, and they make money—lots of money. People like the idea of popping pills to cure diseases. Which is a shame, because the truth is in spite of all the effort put in, we have pitifully few effective antiviral drugs available. Got a cold? Hard luck. And as far as most developing countries are concerned, pricing puts most drugs out of reach of the people who need them. Antiretroviral therapy can keep AIDS patients alive for decades (if you can afford it), but what about the millions who die each year from respiratory infections or diarrhea?

Table 6.8 Antiviral Drugs

Drug	Viruses	Chemical Type	Target
Vidarabine	Herpesviruses	Nucleoside analogue	Virus polymerase
Acyclovir	HSV	Nucleoside analogue	Virus polymerase
Gancyclovir	CMV	Nucleoside analogue	Virus polymerase (requires virus UL98 kinase for activation)
Nucleoside-analogue reverse transcriptase inhibitors (NRTI)— zidovudine (AZT), didanosine (ddl), zalcitabine (ddC), stavudine (d4T), lamivudine (3TC)	Retroviruses (HIV)	Nucleoside analogue	RT
Nonnucleoside reverse transcriptase inhibitors (NNRTI)—nevirapine, delavirdine	Retroviruses: HIV	Nucleoside analogue	RT
Protease inhibitors—saquinavir, ritonavir, indinavir, nelfinavir	HIV	Peptide analogue	HIV protease
Ribavirin	Broad-spectrum: HCV, HSV, measles, mumps, Lassa fever	Triazole carboxamide	RNA mutagen
Amantadine/rimantadine	Influenza A	Tricyclic amine	Matrix protein/ hemagglutinin
Neuraminidase inhibitors—oseltamivir, zanamivir	Influenza A and B	Ethyl ester pro-drug requiring hydrolysis for conversion to the active carboxylate form	Neuraminidase

The alternative to vaccination is to attempt to treat virus infections using drugs that block virus replication (Table 6.8). Historically, the discovery of antiviral drugs was largely down to luck. Spurred on by successes in the treatment of bacterial infections with antibiotics, drug companies launched huge blind-screening programs to identify chemical compounds with antiviral activity, with relatively little success. The key to the success of any antiviral drug lies in its specificity. Almost any stage of virus replication can be a target for a drug, but the drug must be more toxic to the virus than the host. This is measured by the chemotherapeutic index, given by:

$$\frac{\text{Dose of drug that inhibits virus replication}}{\text{Dose of drug that is toxic to host}}$$

The smaller the value of the chemotherapeutic index, the better. In practice, a difference of several orders of magnitude between the two toxicity values is usually required to produce a safe and clinically useful drug. Modern technology, including molecular biology and computer-aided design of chemical compounds, allows the deliberate design of drugs, but it is necessary to "know your enemy"—to understand the key steps in virus replication that might be inhibited. Any of the stages of virus replication can be a target for antiviral intervention. The only requirements are:

- The process targeted must be essential for replication.
- The drug is active against the virus but has "acceptable toxicity" to the host organism.

What degree of toxicity is "acceptable" clearly varies considerably—for example, between a cure for the common cold, which might be sold over the counter and taken by millions of people, and a drug used to treat fatal virus infections such as AIDS.

The attachment phase of replication can be inhibited in two ways, by agents that mimic the VAP and bind to the cellular **receptor** or by agents that mimic the receptor and bind to the VAP. Synthetic peptides are the most logical class of compound to use for this purpose. While this is a promising line of research, there are considerable problems with the clinical use of these substances, primarily the high cost of synthetic peptides and the poor pharmacokinetic properties of many of these synthetic molecules.

It is difficult to target specifically the penetration/uncoating stages of virus replication as relatively little is known about them. Uncoating in particular is largely mediated by cellular enzymes and is therefore a poor target for intervention, although, like penetration, it is often influenced by one or more virus proteins. Amantadine and rimantadine are two drugs that are active against influenza A viruses. The action of these closely related agents is to block cellular membrane ion channels. The target for both drugs is the influenza matrix protein (M_2), but resistance to the drug may also map to the hemagglutinin gene. This biphasic action results from the inability of drug-treated cells to lower the pH of the endosomal compartment (a function normally controlled by the M_2 gene product), which is essential to induce conformational changes in the HA protein to permit membrane fusion (see Chapter 4).

Many viruses have evolved their own specific enzymes to replicate virus nucleic acids preferentially at the expense of cellular molecules. There is often sufficient specificity in virus polymerases to provide a target for an antiviral agent, and this method has produced the majority of the specific antiviral drugs currently in use. The majority of these drugs function as polymerase substrates (i.e., nucleoside/nucleotide) analogues, and their toxicity varies considerably, from some that are well tolerated (e.g., acyclovir) to others that are quite toxic (e.g., azidothymidine or AZT). There is a problem with the pharmacokinetics of these nucleoside analogues in that their typical serum half-life is 1 to 4 hours. Nucleoside analogues are in fact pro-drugs, as they must be phosphorylated before becoming effective—which is key to their selectivity:

- Acyclovir is phosphorylated by HSV thymidine kinase 200 times more efficiently than by cellular enzymes.
- Ganciclovir is 10 times more effective against CMV than acyclovir but must be phosphorylated by a kinase encoded by CMV gene UL97 before it becomes pharmaceutically active.
- Other nucleoside analogues derived from these drugs and active against herpesviruses have been developed (e.g., valciclovir and famciclovir). These compounds have improved pharmacokinetic properties, such as better oral bioavailability and longer half-lives.

In addition to these there are a number of nonnucleoside analogues that inhibit virus polymerases; for example, foscarnet is an analogue of pyrophosphate that interferes with the binding of incoming nucleotide triphosphates by virus DNA polymerases. Ribavirin is a compound with a very wide spectrum of activity against many different viruses, especially against many (−)sense RNA viruses. This drug acts as an RNA mutagen, causing a 10-fold increase in mutagenesis of RNA virus genomes and a 99% loss in virus infectivity after a single round of virus infection in the presence of ribavirin. Ribavirin is thus quite unlike the other nucleoside analogues described above, and its use might become much more widespread in the future if it were not for the frequency of adverse effects associated with this drug.

Virus gene expression is less amenable to chemical intervention than genome replication, because viruses are much more dependent on the cellular machinery for transcription, mRNA splicing, cytoplasmic export, and translation than for replication. To date, no clinically useful drugs that discriminate between virus and cellular gene expression have been developed. As with penetration and uncoating, for the majority of viruses the processes of assembly, maturation, and release are poorly understood and therefore have not yet become targets for antiviral intervention, with the exception of the anti-influenza drugs oseltamivir and zanamivir, which are inhibitors of influenza virus neuraminidase. Neuraminidase is involved in the release of virus particles budding from infected cells, and these drugs are believed to reduce the spread of virus to other cells.

The most striking aspect of antiviral chemotherapy is how few clinically useful drugs are available. As if this were not bad enough, there is also the problem of drug resistance to consider. In practice, the speed and frequency with which resistance arises when drugs are used to treat virus infections varies considerably and depends largely on the biology of the virus involved rather than on the chemistry of the compound. To illustrate this, two extreme cases are described here.

Acyclovir, used to treat HSV infections, is easily the most widely used antiviral drug. This is particularly true in the case of genital herpes, which causes painful recurrent ulcers on the genitals. It is estimated that 40 to 60 million people suffer from this condition in the United States. Fortunately, resistance to acyclovir arises infrequently. This is partly due to the high fidelity with which the DNA genome of HSV is copied (Chapter 3). Mechanisms that give rise to acyclovir resistance include:

- HSV *pol* gene mutants that do not incorporate acyclovir
- HSV thymidine kinase (TK) mutants in which TK activity is absent (TK$^-$) or reduced or shows altered substrate specificity

Strangely, it is possible to find mutations that give rise to each of these phenotypes with a frequency of 1×10^{-3} to 1×10^{-4} in clinical HSV isolates. The discrepancy between this and the very low frequency with which resistance is recorded clinically is probably explained by the observation that most *pol*/TK mutants appear to be attenuated (e.g., TK$^-$ mutants of HSV do not reactivate from the latent state).

Conversely, AZT treatment of HIV infection is much less effective. In untreated HIV-infected individuals, AZT produces a rise in the numbers of CD4$^+$ cells within 2–6 weeks. However, this beneficial effect is transient; after 20 weeks, CD4$^+$ T-cell counts generally revert to baseline. This is due partly to the development of AZT resistance in treated HIV populations and to the toxicity of AZT on hematopoiesis, as the chemotherapeutic index of AZT is much worse than that of acyclovir. AZT resistance is initiated by the acquisition of a mutation in the HIV reverse transcriptase (RT) gene at codon 215. In conjunction with two to three additional mutations in the RT gene, a fully AZT-resistant phenotype develops. After 20 weeks of treatment, 40–50% of AZT-treated patients develop at least one of these mutations. This high frequency is due to the error-prone nature of reverse transcription (Chapter 3).

Because of the large number of replicating HIV genomes in infected patients (Chapter 7), many mistakes occur continuously. It has been shown that the mutations that confer resistance already exist in untreated virus populations. Thus, treatment with AZT does not cause but merely selects these resistant viruses from the total pool. With other anti-RT drugs, such as didanosine (ddI), a resistant phenotype can result from a single base pair change, but ddI has an even lower therapeutic index than AZT, and relatively low levels of resistance can potentially render this drug useless. However, some combinations of resistant mutations may make it difficult for HIV to replicate, and resistance to one RT inhibitor may counteract resistance to another. The current strategy for therapy of HIV infection is known as HAART (highly active antiretroviral therapy) and employs combinations of different drugs such as a protease inhibitor plus two nucleoside RT inhibitors. Molecular mechanisms of resistance and drug interactions are both important to consider when designing combination regimes:

- Combinations such as AZT + ddI or AZT + 3TC have antagonistic patterns of resistance and are effective.
- Combinations such as ddC + 3TC that show cross-reactive resistance should be avoided.

Certain protease inhibitors affect liver function and can favorably affect the pharmacokinetics of RT inhibitors taken in combination. Other potential benefits of combination antiviral therapy include lower toxicity profiles and the use of drugs that may have different tissue distributions or cell tropisms. Combination therapy may also prevent or delay the development of drug resistance. Combinations of drugs that can be employed include not only small synthetic molecules but also "biological response modifiers" such as interleukins and IFNs.

SUMMARY

Virus infection is a complex, multistage interaction between the virus and the host organism. The course and eventual outcome of any infection are the result of a balance between host and virus processes. Host factors involved include exposure to different routes of virus transmission and the control of virus replication by the immune response. Virus processes include the initial infection of the host, spread throughout the host, and regulation of gene expression to evade the immune response. Medical intervention against virus infections includes the use of vaccines to stimulate the immune response and drugs to inhibit virus replication. Molecular biology is stimulating the production of a new generation of antiviral drugs and vaccines.

Further Reading

Aliyari, R., Ding, S.W., 2009. RNA-based viral immunity initiated by the Dicer family of host immune receptors. Immunol. Rev. 227 (1), 176–188.

Best, S.M., 2008. Viral subversion of apoptotic enzymes: escape from death row. Annu. Rev. Microbiol. 62, 171–192.

Coiras, M., López-Huertas, M.R., Pérez-Olmeda, M., Alcamí, J., 2009. Understanding HIV-1 latency provides clues for the eradication of long-term reservoirs. Nat. Rev. Microbiol. 7 (11), 798–812. Available from: http://dx.doi.org/10.1038/nrmicro2223.

Crotty, S., Andino, R., 2002. Implications of high RNA virus mutation rates: lethal mutagenesis and the antiviral drug ribavirin. Microbes Infect. 4, 1301–1307.

Cullen, B.R., 2010. Five questions about viruses and microRNAs. PLoS Pathog. 6 (2), e1000787. Available from: http://dx.doi.org/10.1371/journal.ppat.1000787.

Cullen, B.R., 2013. How do viruses avoid inhibition by endogenous cellular microRNAs? PLoS Pathog. 9 (11), e1003694. Available from: http://dx.doi.org/10.1371/journal.ppat.1003694.

Ferraro, B., Morrow, M.P., Hutnick, N.A., Shin, T.H., Lucke, C.E., Weiner, D.B., 2011. Clinical applications of DNA vaccines: current progress. Clin. Infect. Dis. 53 (3), 296–302.

Lisnića, V.J., Krmpotića, A., Jonjića, S., 2010. Modulation of natural killer cell activity by viruses. Curr. Opin. Microbiol. 13 (4), 530–539. Available from: http://dx.doi.org/10.1016/j.mib.2010.05.011.

Loc-Carrillo, C., Abedon, S.T., 2011. Pros and cons of phage therapy. Bacteriophage 1 (2), 111–114. Available from: http://dx.doi.org/10.4161/bact.1.2.14590.

Lu, L.F., Liston, A., 2009. MicroRNA in the immune system, microRNA as an immune system. Immunology 127 (3), 291–298.

Marraffini, L.A., Sontheimer, E.J., 2010. CRISPR interference: RNA-directed adaptive immunity in bacteria and archaea. Nat. Rev. Genet. 11 (3), 181–190.

Miest, T.S., Cattaneo, R., 2014. New viruses for cancer therapy: meeting clinical needs. Nat. Rev. Microbiol. 12 (1), 23–34.

Owen, J., Punt, J., Stranford, S., 2013. Kuby Immunology. seventh ed. W.H. Freeman, ISBN 1464137846.

Pallas, V., García, J.A., 2011. How do plant viruses induce disease? Interactions and interference with host components. J. Gen. Virol. 92 (12), 2691–2705.

Randall, R.E., Goodbourn, S., 2008. Interferons and viruses: an interplay between induction, signalling, antiviral responses and virus countermeasures. J. Gen. Virol. 89 (1), 1–47.

Razonable, R.R., 2011. Antiviral drugs for viruses other than human immunodeficiency virus. Mayo Clin. Proc. 86 (10), 1009–1026.

Roossinck, M.J., 2013. Plant virus ecology. PLoS Pathog. 9 (5), e1003304. Available from: http://dx.doi.org/10.1371/journal.ppat.1003304.

Subbarao, K., Matsuoka, Y., 2013. The prospects and challenges of universal vaccines for influenza. Trends Microbiol. 21 (7), 350–358.

Tortorella, D., et al., 2000. Viral subversion of the immune system. Annu. Rev. Immunol. 18, 861–926.

Versteeg, G.A., García-Sastre, A., 2010. Viral tricks to grid-lock the type I interferon system. Curr. Opin. Microbiol. 13 (4), 508–516.

Woller, N., Gürlevik, E., Ureche, C.I., Schumacher, A., Kühnel, F., 2014. Oncolytic viruses as anticancer vaccines. Front. Oncol. 4, 188. Available from: http://dx.doi.org/10.3389/fonc.2014.00188.

CHAPTER 7

Pathogenesis

Intended Learning Outcomes

On completing this chapter you should be able to:

- Discuss the link between virus infection and disease.
- Explain how virus infection may injure the body, including how HIV infection causes AIDS and how some viruses may cause cancer.
- Define what emerging viruses are and predict what the future may hold for us.

Pathogenicity, the capacity of one organism to cause disease in another, is a complex and variable process. At the simplest level there is the question of defining what disease is. An all-embracing definition would be that disease is a departure from the normal physiological parameters of an organism. This could range from a temporary and very minor condition, such as a slightly raised temperature or lack of energy, to chronic pathologic conditions that eventually result in death. Any of these conditions may result from a tremendous number of internal or external sources. There is rarely one single factor that "causes" a disease; most disease states are multifactorial at some level.

In considering virus diseases, two aspects are involved—the direct effects of virus replication and the effects of body's responses to the infection. The course of any virus infection is determined by a delicate and dynamic balance between the host and the virus, as described in Chapter 6. The extent and severity of virus pathogenesis is determined similarly. In some virus infections, most of the pathologic symptoms observed are not directly caused by virus replication but are to the side effects of the immune response. Inflammation, fever, headaches, and skin rashes are not usually caused by viruses themselves but by the cells of the immune system due to the release of potent chemicals such as interferons and interleukins. In the most extreme cases, it is possible that none of the pathologic effects of certain diseases is caused directly by the virus, except that its presence stimulates activation of the immune system.

CONTENTS

Principles of Molecular Virology. DOI: http://dx.doi.org/10.1016/B978-0-12-801946-7.00007-9

In the past few decades, molecular analysis has contributed enormously to our understanding of virus pathogenesis. Nucleotide sequencing and site-directed mutagenesis have been used to explore molecular determinants of virulence in many different viruses. Specific sequences and structures found only in disease-causing strains of virus and not in closely related attenuated or avirulent strains have been identified. Sequence analysis has also led to the identification of T-cell and B-cell epitopes on virus proteins responsible for their recognition by the immune system. Unfortunately, these advances do not automatically lead to an understanding of the mechanisms responsible for pathogenicity.

Unlike the rest of this book, this chapter is specifically about viruses that cause disease in animals. It does not discuss viruses that cause disease in plants since this has already been considered in Chapter 6. Three major aspects of virus pathogenesis are considered: direct cell damage resulting from virus replication, damage resulting from immune activation or suppression, and cell transformation caused by viruses.

BOX 7.1 DON'T BLAME THE VIRUSES!

Virus pathogenesis is an abnormal and fairly rare situation. The majority of virus infections are silent and do not result in any outward signs of disease. It is sometimes said that viruses would disappear if they killed their hosts. This is not necessarily true. It is possible to imagine viruses with a hit-and-run strategy, moving quickly from one dying host to the next and relying on continuing circulation for their survival. Nevertheless, there is a clear tendency for viruses not to injure their hosts if possible. A good example of this is the rabies virus. The symptoms of human rabies virus infections are truly dreadful, but thankfully rare. In its normal hosts (e.g., foxes), rabies virus infection produces a much milder disease that does not usually kill the animal. Humans are an unnatural,

dead-end host for this virus, and the severity of human rabies is as extreme as the condition is rare. Ideally, a virus would not even provoke an immune response from its host, or at least would be able to hide to avoid the effects. Herpesviruses and some retroviruses have evolved complex lifestyles that enable them to get close to this objective, remaining silent for much of the time. Of course, fatal infections such as rabies and acquired immune deficiency syndrome (AIDS) always grab the headlines. Much less effort has been devoted to isolating and studying the many viruses that have not (yet) caused well-defined diseases in humans, domestic animals, or economically valuable crop plants.

MECHANISMS OF CELLULAR INJURY

Virus infection often results in a number of changes that are detectable by visual or biochemical examination of infected cells. These changes result from the production of virus proteins and nucleic acids, but also from alterations to the biosynthetic capabilities of infected cells. Virus replication sequesters cellular apparatus such as ribosomes and raw materials that would normally be devoted to synthesizing molecules required by the cell. Eukaryotic cells must carry out constant macromolecular synthesis, whether they are growing and dividing or in a state of quiescence. A growing cell needs to manufacture more proteins, more nucleic acids, and more of all of its components to increase its size before dividing. However, there is another reason for such continuous activity. The function of all cells is regulated by controlled expression of their genetic information and the subsequent degradation of the molecules produced. Such control relies on a delicate and dynamic balance between synthesis and decay which determines the intracellular levels of all the important molecules in the cell. This is particularly true of the control of the cell cycle, which determines the behavior of dividing cells (see "Cell Transformation by DNA Viruses," below). In general terms, a number of common phenotypic changes can be recognized in virus-infected cells. These changes are often referred to as the cytopathic effects (c.p.e.) of a virus, and include:

- **Altered shape:** Adherent cells that are normally attached to other cells (*in vivo*) or an artificial substrate (*in vitro*) may assume a rounded shape different from their normal flattened appearance. The extended "processes" (extensions of the cell surface resembling tendrils) involved in attachment or mobility are withdrawn into the cell.
- **Detachment from the substrate:** For adherent cells, this is the stage of cell damage that follows that above. Both of these effects are caused by partial degradation or disruption of the cytoskeleton that is normally responsible for maintaining the shape of the cell.
- **Lysis:** This is the most extreme case, where the entire cell breaks down. Membrane integrity is lost, and the cell may swell due to the absorption of extracellular fluid and finally break open. This is an extreme case of cell damage, and it is important to realize that not all viruses induce this effect, although they may cause other cytopathic effects. Lysis is beneficial to a virus in that it provides an obvious method of releasing new virus particles from an infected cell; however, there are alternative ways of achieving this, such as release by budding (Chapter 4).
- **Membrane fusion:** The membranes of adjacent cells fuse, resulting in a mass of cytoplasm containing more than one nucleus, known as a syncytium, or, depending on the number of cells that merge, a giant cell. Fused cells are short lived and subsequently lyse—apart from direct effects of the virus, they cannot tolerate more than one nonsynchronized nucleus per cell.
- **Membrane permeability:** A number of viruses cause an increase in membrane permeability, allowing an influx of extracellular ions such as sodium. Translation of some virus mRNAs is resistant to high concentrations of sodium ions, permitting the expression of virus genes at the expense of cellular messages.

- **Inclusion bodies:** These are areas of the cell where virus components have accumulated. They are frequent sites of virus assembly, and some cellular inclusions consist of crystalline arrays of virus particles. It is not clear how these structures damage the cell, but they are frequently associated with viruses that cause cell lysis, such as herpesviruses and rabies virus.
- **Apoptosis:** Virus infection may trigger apoptosis ("programmed cell death"), a highly specific mechanism involved in the normal growth and development of organisms (see Chapter 6).

In some cases, a great deal of detail is known about the molecular mechanisms of cell injury. A number of viruses that cause cell lysis exhibit a phenomenon known as shutoff early in infection. Shutoff is the sudden and dramatic cessation of most host-cell macromolecular synthesis. In poliovirus-infected cells, shutoff is the result of production of the virus 2A protein. This molecule is a protease that cleaves the p220 component of eIF-4F, a complex of proteins required for cap-dependent translation of messenger RNAs by ribosomes. Because poliovirus RNA does not have a 5' methylated cap but is modified by the addition of the VPg protein, virus RNA continues to be translated. In poliovirus-infected cells, the dissociation of mRNAs and polyribosomes from the cytoskeleton can be observed, and this is the reason for the inability of the cell to translate its own messages. A few hours after translation ceases, lysis of the cell occurs.

In other cases, cessation of cellular macromolecular synthesis results from a different molecular mechanism. For many viruses, the sequence of events that occurs is not known. In the case of adenoviruses, the penton protein (part of the virus capsid) has a toxic effect on cells. Although its precise action on cells is not known, addition of purified penton protein to cultured cells results in their rapid death. Toxin production by pathogenic bacteria is a common phenomenon, but this is the only well-established case of a virus-encoded molecule with a toxin-like action. However, some of the normal contents of cells released on lysis may have toxic effects on other cells, and antigens that are not recognized as "self" by the body (e.g., nuclear proteins) may result in immune activation and inflammation. The adenovirus E3−11.6K protein is synthesized in small amounts from the E3 promoter at early stages of infection and in large amounts from the major late promoter at late stages of infection (Chapter 5). It has recently been shown that E3−11.6K is required for the lysis of adenovirus-infected cells and the release of virus particles from the nucleus.

Membrane fusion is the result of virus-encoded proteins required for infection of cells (see Chapter 4), typically, the glycoproteins of enveloped viruses. One of the best-known examples of such a protein comes from Sendai virus (a paramyxovirus), which has been used to induce cell fusion during the production of monoclonal antibodies (Chapter 1). At least 9 of the 11 known herpes simplex virus (HSV/HHV-1) glycoproteins have been characterized regarding their role in virus replication. Several of these proteins are involved in fusion of the virus envelope with the cell membrane and also in cell penetration. Production of fused syncytia is a common feature of HSV infection.

Another virus that causes cell fusion is human immunodeficiency virus (HIV). Infection of $CD4^+$ cells with some but not all isolates of HIV causes cell–cell fusion and the production of syncytia or giant cells (Figure 7.1). The protein responsible for this is the transmembrane envelope glycoprotein of the virus (gp41), and the domain near the amino-terminus responsible for this fusogenic activity has been identified by molecular genetic analysis. Because HIV infects $CD4^+$ cells and it is the reduction in the number of these crucial cells of the immune system that is the most obvious defect in AIDS, it was initially believed that direct killing of these cells by the virus was the basis for the pathogenesis of AIDS. Although direct cell killing by HIV undoubtedly occurs *in vivo*, it is now believed that the pathogenesis of AIDS is considerably more complex (see "Viruses and Immunodeficiency," below). Many animal retroviruses also cause cell killing and, in most cases, it appears that the envelope protein of the virus is required, although there may be more than one mechanism involved.

VIRUSES AND IMMUNODEFICIENCY

At least two groups of viruses, herpesviruses and retroviruses, directly infect the cells of the immune system. This has important consequences for the outcome of the infection and for the immune system of the host. HSV establishes a systemic infection, spreading via the bloodstream in association with platelets, but it does not show particular tropism for cells of the immune system. However, *Herpes saimirii* and Marek's disease virus are herpesviruses that cause lymphoproliferative diseases (but not clonal tumors) in monkeys and chickens, respectively. The most recently discovered human herpesviruses (HHVs)—HHV-6, HHV-7, and HHV-8—all infect lymphocytes (Chapter 8).

Epstein–Barr virus (EBV; HHV-4) infection of B-cells leads to their immortalization and proliferation, resulting in "glandular fever" or mononucleosis, a debilitating but benign condition. EBV was first identified in a lymphoblastoid cell line derived from Burkitt's lymphoma and, in rare instances, EBV infection may lead to the formation of a malignant tumor (see "Cell Transformation by DNA Viruses," below). While some herpesviruses such as HSV are highly cytopathic, most of the lymphotropic herpesviruses do not cause a significant degree of cellular injury. However, infection of the delicate

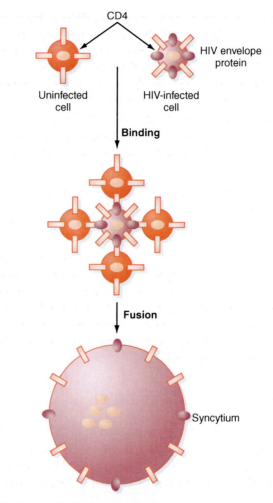

FIGURE 7.1 Mechanism of HIV-induced cell fusion.
The virus envelope glycoprotein, which plays a role on virus particles in receptor binding and membrane fusion, is expressed on the surface of infected cells. Uninfected CD4$^+$ cells coming into contact with these infected cells are fused together to form a multinucleate syncytium.

cells of the immune system may perturb their normal function. Because the immune system is internally regulated by complex networks of interlinking signals, relatively small changes in cellular function can result in its collapse. Alteration of the normal pattern of production of cytokines could have profound effects on immune function. The *trans*-regulatory proteins involved in the control of herpesvirus gene expression may also affect the transcription of cellular genes; therefore, the effects of herpesviruses on immune cells are more complex than just cell killing.

Retroviruses cause a variety of pathogenic conditions including paralysis, arthritis, anemia, and malignant cellular transformation. A significant number of retroviruses infect the cells of the immune system. Although these infections may lead to a diverse array of diseases and hematopoetic abnormalities such as anemia and lymphoproliferation, the most commonly recognized consequence of retrovirus infection is the formation of lymphoid tumors. However, some degree of immunodeficiency, ranging from very mild to quite severe, is a common consequence of the interference with the immune system resulting from the presence of a lymphoid or myeloid tumor.

The most prominent aspect of virus-induced immunodeficiency is acquired immunodeficiency syndrome (AIDS), a consequence of infection with HIV, a member of the genus *Lentivirus* of the *Retroviridae*. A number of similar lentiviruses cause immunodeficiency diseases in animals. Unlike infection by other types of retrovirus, HIV infection does not directly result in the formation of tumors. Some tumors such as B-cell lymphomas are sometimes seen in AIDS patients, but these are a consequence of the lack of immune surveillance that is responsible for the destruction of tumors in healthy individuals. The clinical course of AIDS is long and very variable. A great number of different abnormalities of the immune system are seen in AIDS. As a result of the biology of lentivirus infections, the pathogenesis of AIDS is highly complex (Figure 7.2).

It is still not clear how much of the pathology of AIDS is caused directly by the virus and how much is caused by the immune system. Numerous models have been suggested to explain how HIV causes immunodeficiency. These mechanisms are not mutually exclusive and indeed it is probable that the underlying loss of CD4$^+$ cells (see Chapter 6) in AIDS is complex and multifactorial. AIDS is defined as the presence of HIV infection, plus one or both of the following:

- A CD4$^+$ T-cell count of less than 200 cells/ml of blood (the normal count is 600−1,000 per ml).
- Development of an opportunistic infection that occurs when the immune system is not working correctly, such as *Pneumocystis carinii* pneumonia (PCP), certain eye diseases, encephalitis, and some specific tumors such as Kaposi's sarcoma (KS).

The best way to avoid AIDS is not to become infected with HIV, but that is not much help to the 35 million people worldwide who already are infected with the virus. If we are to find a cure for AIDS, we need to understand the mechanisms by which the virus causes the disease. Although the basic biology of HIV is well understood (see Further Reading at the end of this chapter), scientists have never had a complete understanding of the processes by

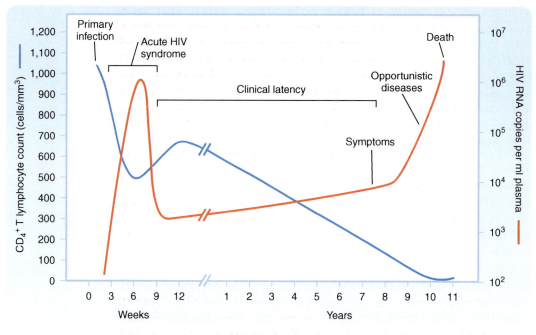

FIGURE 7.2 Time course of HIV infection.
This diagram shows a typical sequence of events in an HIV-infected person during the interval between infection with the virus and the development of AIDS.

which CD4$^+$ T-helper cells are depleted in HIV infection, and therefore have never been able to fully explain why HIV destroys the body's supply of these vital cells.

There have been many theories about how HIV infection results in AIDS. Soon after HIV was discovered in the 1980s it was shown that the virus could kill CD4$^+$ cells in culture. Early experiments suggested there might not be enough virus present in AIDS patients to account for all the cell loss seen. Later, sensitive PCR techniques suggested that with the amount of virus present in infected individuals, the CD4$^+$ cell count should in fact decline much faster and AIDS develop much earlier than it does after HIV infection. Researchers have used a "tap and drain" analogy to describe CD4$^+$ cell loss in HIV infection. In this description of the disease, CD4$^+$ cells (like water in a sink) are constantly being eliminated by HIV (the drain), while the body is constantly replacing them with new ones (the tap). Over time, the tap cannot keep up with the drain, and CD4 counts begin to drop, leaving the body susceptible to the infections that define AIDS. CD4$^+$ cells that are activated in response to invading microbes (including HIV itself) are highly susceptible to infection with the virus, and following infection these cells may produce many new copies of HIV before dying. One explanation for CD4$^+$ cell loss is the "runaway" hypothesis, in which CD4$^+$ cells infected by HIV produce more virus particles, which activate more CD4$^+$ cells that in turn become infected, leading to a positive feedback cycle of CD4$^+$ cell activation, infection, HIV production, and cell destruction. Unfortunately, mathematical models consisting of a series of equations to describe the processes by which CD4$^+$ cells are produced and eliminated suggest that if the "runaway" hypothesis was correct, then CD4$^+$ cells in HIV-infected individuals would fall to low levels over a few months, not over several years as usually happens. This implies that the "runaway" hypothesis cannot explain the slow pace of CD4$^+$ cell depletion in HIV infection. That leaves open the question of what exactly is going on between the time someone becomes infected with HIV and the time that they develop AIDS. While virus adaptation (antigenic variation) is important in the biology of HIV, this alone cannot explain the whole story.

In general, HIV is regarded as an incurable infection, although in many cases doctors are able to stave off the onset of AIDS by giving patients sustained courses of antiretrovirus drugs. As a retrovirus, the biology of HIV, including integration of the virus genome into host-cell chromosomes, is a major problem in eradicating the virus from the body (see Chapter 3). In HIV-infected people receiving antiviral therapy there is a reservoir of latently infected resting CD4$^+$ T cells. Many HIV patients can manage their infection with a cocktails of antiretrovirus drugs which can reduce their "viral load"—the amount of virus circulating in the blood plasma—to undetectable levels. But even in such "noninfectious" patients HIV is still lurking in gut tissues, and still infecting other immune cells in the blood. Mathematical modeling and clinical observations suggest it might not ever be possible to completely eradicate the virus from the body with current therapies. The hope is that new approaches such as RNAi (see Chapter 6) might one day be able to tackle this latent virus pool and completely eliminate the virus from the body, curing the infection. Recent work has focused on the development of therapies to disrupt virus latency and expose the virus to replication-blocking drugs and to the immune response which exists in HIV-positive people. Even if this is possible, the cost of these advanced therapies would be beyond the reach of developing countries where the majority of HIV-infected people live. It may be that adaptation of HIV to replication in humans may eventually temper the pathogenicity of the virus by lowering its replication capacity. However, this process will be speeded up by antiviral therapies in many different forms, including genetic, drugs, and vaccines.

BOX 7.2 STEALTHY DOES IT

The more we study viruses the more examples we find of viruses interacting with the immune system. Not interacting as in "Argh! I'm dead", but interacting as in "I wonder what happens if I twist this knob?". Almost all viruses moderate the immune responses directed against them. This makes sense—if they couldn't do this, they probably wouldn't be able to replicate. And some viruses are masters of the art, subtly tweaking and muting strands of the immune system to make life easier for them. Herpesviruses and poxviruses spring to mind. In comparison to them, viruses which go for an all-out assault on the body or on the immune system seem like amateurs. The consequences on their hosts are devastating, which is bad for both the virus and the host. So let's hear it for the true masters of the craft of sneaking around, of getting on with things quietly.

VIRUS-RELATED DISEASES

Virus infections are believed to be a necessary prerequisite for a number of human diseases which are not directly caused by the virus. In some instances, the link between a particular virus and a pathological condition is well established, but it is clear that the pathogenesis of the disease is complex and also involves the immune system of the host. In other cases, the pathogenic involvement of a particular virus is less certain and, in a few instances, rather speculative.

Although the incidence of measles virus infection has been reduced sharply by vaccination (Chapter 6), measles still causes thousands of deaths worldwide each year. The normal course of measles virus infection is an acute febrile illness during which the virus spreads throughout the body, infecting many tissues. The vast majority of people spontaneously recover from the disease without any lasting harm. In rare cases (about 1 in 2,000), measles may progress to a severe encephalitis. This is still an acute condition that either regresses or kills the patient within a few weeks; however, there is another, much rarer late consequence of measles virus infection that occurs many months or years after initial infection of the host. This is the condition known as subacute sclerosing panencephalitis (SSPE). Evidence of prior measles virus infection (antibodies or direct detection of the virus) is found in all patients with SSPE, whether they can recall having a symptomatic case of measles or not. In about 1 in 300,000 cases of measles, the virus is not cleared from the body by the immune system but establishes a persistent infection in the CNS. In this condition, virus replication continues at a low level, but defects in the envelope protein genes prevent the production of extracellular infectious virus particles. The lack of envelope protein production causes the failure of the immune system to recognize and eliminate infected cells; however, the virus is able to spread directly from cell to cell, bypassing the usual route of infection. It is not known to what extent damage to the cells of the brain is caused directly by virus replication or whether there is any contribution by the immune system to the pathogenesis of SSPE. Vaccination against measles virus and the prevention of primary infection should ultimately eliminate this condition.

Another well-established case where the immune system is implicated in pathogenesis concerns dengue virus infections. Dengue virus is a flavivirus that is transmitted from one human host to another via mosquitoes. The primary infection may be asymptomatic or may result in dengue fever. Dengue fever is normally a self-limited illness from which patients recover after 7–10 days without further complications. Following primary infections, patients carry antibodies to the virus. Unfortunately, there are four serotypes of dengue virus (DEN-1, 2, 3, and 4), and the presence of antibody directed against one type does not give cross-protection against the other three; worse still is the fact that antibodies can enhance the infection of peripheral blood mononuclear cells by Fc-receptor-mediated uptake of antibody-coated dengue virus particles (see Chapter 4). In a few cases, the consequences of dengue virus infection are much more severe than the usual fever. Dengue hemorrhagic fever (DHF) is a life-threatening disease. In the most extreme cases, so much internal hemorrhaging occurs that hypovolemic shock (dengue shock syndrome or DSS) occurs. DSS is frequently fatal. The cause of shock in dengue and other hemorrhagic fevers is partly due to the virus, but largely due to immune-mediated damage of virus-infected cells (Figure 7.3). DHF and DSS following primary dengue virus infections occur in approximately 1 in 14,000 and 1 in 500 patients, respectively; however, after secondary dengue virus infections, the incidence of DHF is 1 in 90 and DSS 1 in 50, as cross-reactive but nonneutralizing antibodies to the virus are now present. These figures show the problems of cross-infection with different serotypes of dengue virus, and the difficulties that must be faced in developing a safe vaccine against the virus. Dengue virus is discussed further later in this chapter (see "New and Emergent Viruses").

Another instance where virus vaccines have resulted in increased pathology rather than the prevention of disease is the occurrence of postvaccination Reye's syndrome. Reye's syndrome is a neurological condition involving acute cerebral edema and occurs almost exclusively in children. It is well known as a rare postinfection complication of a number of different viruses, but most commonly influenza virus and Varicella–Zoster virus (chicken pox). Symptoms include frequent vomiting, painful headaches, behavioral changes, extreme tiredness, and disorientation. The chances of contracting Reye's syndrome are increased if aspirin is administered during the initial illness. The basis for the pathogenesis of this condition is completely unknown, but some of the most unfortunate cases have followed the administration of experimental influenza virus vaccines.

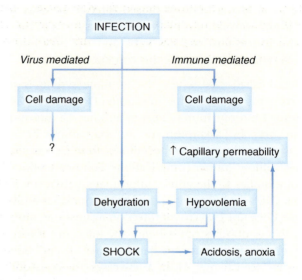

FIGURE 7.3 Causes of shock in hemorrhagic fevers.
The cause of hypovolemic shock in dengue and other hemorrhagic fevers is partly due to the virus, but largely due to immune-mediated damage of virus-infected cells.

Guillain–Barré syndrome is another mysterious condition in which demyelination of nerves results in partial paralysis and muscle weakness. The onset of Guillain–Barré syndrome usually follows an acute "virus-like" infection, but no single agent has ever been firmly associated with this condition. Kawasaki disease is similar to Reye's syndrome in that it occurs in children but is distinct in that it results in serious damage to the heart. Like Guillain–Barré syndrome, Kawasaki disease appears to follow acute infections. The disease itself is not infectious but does appear to occur in epidemics, which suggests an infectious agent as the cause. A large number of bacterial and virus pathogens have been suggested to be associated with the induction of Kawasaki disease, but once again the underlying cause of the pathology is unknown. It would appear that acute infection itself rather than a particular pathogen may be responsible for the onset of these diseases.

In recent years, there has been a search for an agent responsible for a newly diagnosed disease called chronic fatigue syndrome (CFS) or myalgic encephalomyelitis. Unlike the other conditions described above, CFS is a rather ill-defined disease and is not recognized by all physicians. Recent research has discounted the earlier idea that EBV might cause CFS, but a variety of other possible virus causes, including other herpesviruses, enteroviruses, and retroviruses, have also been suggested. In October 2009 it was reported that a novel retrovirus, xenotropic murine leukemia virus-related virus (XMRV), might be a possible cause. Unfortunately, subsequent research findings about XMRV proved to be contradictory and confusing, and there is no strong evidence for the role of XMRV in CFS.

BACTERIOPHAGES AND HUMAN DISEASE

Can bacteriophages, viruses that are only capable of infecting prokaryotic cells, play a role in human disease? Surprisingly, the answer is yes. Shiga toxin (Stx)-producing *Escherichia coli* (STEC) are able to cause intestinal foodborne diseases such as diarrhea and hemorrhagic colitis. STEC serotype O157:H7, the "hamburger bug," has received much attention in recent years. STEC infections can lead to fatal complications, such as hemolytic−uremic syndrome, as well as neurological disorders. The major virulence characteristics of these strains of bacteria are the ability to colonize the bowel (a natural trait of *E. coli*) and the production of secreted "Shiga toxins," which can damage endothelial and tubular cells and may result in acute kidney failure. At least 100 different *E. coli* serotypes produce Stx toxins, and STEC bacteria occur frequently in the bowels of cattle and other domestic animals such as sheep, goats, pigs, and horses. Meat is infected by fecal contamination, usually at the time of slaughter. Ground meat such as hamburger is particularly dangerous as surface bacterial contamination may become buried deep within the meat where it may not be inactivated by cooking.

What has this got to do with bacteriophages? Various types of Stx are known, but they fall into two main types: Stx1 and Stx2. The Stx1 and Stx2 toxin genes are encoded in the genome of lysogenic lambda-like prophages within the bacteria. Stimuli such as UV light or mitomycin C are known to induce these prophages to release a crop of phage particles which can infect and lysogenize other susceptible bacteria within the gut, accounting for the high prevalence of STEC bacteria (up to 50% of cattle in some herds). Recent research has shown that the scandalous overuse of antibiotics as "growth promoters" in animal husbandry and even antibiotic treatment of infected people can stimulate the production of phage particles and contributes to the increased prevalence of STEC bacteria and growing human death toll. Other bacterial virulence determinants are also encoded by lysogenic phages (e.g., diphtheria toxin, *Streptococcus* erythrogenic toxins, *Staphylococcus* enterotoxins), although the selective pressures that maintain these arrangements are not yet understood. Emerging bacterial genome sequence data strongly indicate that phages have been responsible for spreading virulence determinants across a wide range of pathogens.

The other area where bacteriophages may influence human illness is phage therapy—the use of bacteriophages as antibiotics. This is not a new idea, with initial experiments having been performed (unsuccessfully) shortly after the discovery of bacteriophages almost 100 years ago (Appendix 3); however, with increasing resistance of bacteria to antibiotics and the emergence of "superbugs" immune to all effective treatments, this idea has experienced a resurgence of interest. Although attractive in theory, this approach suffers from a number of defects:

- Bacteriophages are quite specific in their receptor usage and hence the strains of bacteria they can infect; thus, they are "narrow spectrum" antibacterial agents.
- Bacteria exposed to bacteriophages rapidly develop resistance to infection by downregulating or mutating the phage receptor.
- Liberation of endotoxin as a consequence of widespread lysis of bacteria within the body can lead to toxic shock.
- Repeated administration of bacteriophages results in an immune response that neutralizes the phage particles before they can act.

It may be, however, that this is a useful therapy for certain bacterial infections that cannot be treated by conventional means. Recently, it has been shown that bioengineered antibodies can be delivered to the brain by bacteriophage vectors, and this novel approach is being investigated for the treatment of Alzheimer's disease and cocaine addiction.

CELL TRANSFORMATION BY VIRUSES

Transformation is a change in the morphological, biochemical, or growth parameters of a cell. Transformation may or may not result in cells able to produce tumors in experimental animals, which is properly known as neoplastic transformation; therefore, transformed cells do not automatically result in the development of "cancer." Carcinogenesis (or more properly, oncogenesis) is a complex, multistep process in which cellular transformation may be only the first, although essential, step along the way. Transformed cells have an altered phenotype, which is displayed as one (or more) of the following characteristics:

- **Loss of anchorage dependence:** Normal (i.e., nontransformed) adherent cells such as fibroblasts or epithelial cells require a surface to which they can adhere. In the body, this requirement is supplied by adjacent cells or structures; *in vitro*, it is met by the glass or plastic vessels in which the cells are cultivated. Some transformed cells lose the ability to adhere to solid surfaces and float free (or in clumps) in the culture medium without loss of viability.

- **Loss of contact inhibition:** Normal adherent cells in culture divide and grow until they have coated all the available surface for attachment. At this point, when adjacent cells are touching each other, cell division stops—the cells do not continue to grow and pile up on top of one another. Many transformed cells have lost this characteristic. Single transformed cell in a culture dish becomes visible as small thickened areas of growth called "transformed foci"—clones of cells all derived from a single original cell.
- **Colony formation in semisolid media:** Most normal cells (both adherent and nonadherent cells such as lymphocytes) will not grow in media that are partially solid due to the addition of substances such as agarose or hydroxymethyl cellulose; however, many transformed cells will grow under these conditions, forming colonies since movement of the cells is restricted by the medium.
- **Decreased requirements for growth factors:** All cells require multiple factors for growth. In a broad sense, these include compounds such as ions, vitamins, and hormones that cannot be manufactured by the cell. More specifically, it includes regulatory peptides such as epidermal growth factor and platelet-derived growth factor that regulate the growth of cells. These are potent molecules that have powerful effects on cell growth. Some transformed cells may have decreased or may even have lost their requirement for particular factors. The production by a cell of a growth factor required for its own growth is known as autocrine stimulation and is one route by which cells may be transformed.

Cell transformation is a single-hit process; that is, a single virus transforms a single cell (c.f. oncogenesis, which is the formation of tumors and a multi-step process). All or part of the virus genome persists in the transformed cell and is usually (but not always) integrated into the host-cell chromatin. Transformation is usually accompanied by continued expression of a limited repertoire of virus genes or rarely by productive infection. Virus genomes found in transformed cells are frequently replication defective and contain substantial deletions.

Transformation is mediated by proteins encoded by oncogenes. These regulatory genes can be grouped in several ways—for example, by their origins, biochemical function, or subcellular locations (Table 7.1). Cell-transforming viruses may have RNA or DNA genomes, but all have at least a DNA stage in their replication cycle; that is, the only RNA viruses directly capable of cell transformation are the retroviruses (Table 7.2). Certain retroviruses carry homologues of c-*oncs* derived originally from the cellular genes and known as v-*oncs*. In contrast, the oncogenes of cell-transforming DNA viruses are unique to the virus genome—there are no homologous sequences present in

Table 7.1 Categories of Oncogenes

Type	Example
Extracellular growth factors (homologues of normal growth factors)	c-sis: Encodes the PDGF B chain (v-sis in simian sarcoma virus)
	int-2: Encodes a fibroblast growth factor (FGF)-related growth factor (common site of integration for mouse MMTV)
Receptor tyrosine kinases (associated with the inner surface of the cell membrane)	c-fms: Encodes the colony-stimulating factor 1 (CSF-1) receptor—first identified as a retrovirus oncogene
	c-kit: Encodes the mast cell growth factor receptor
Membrane-associated nonreceptor tyrosine kinases (signal transduction)	c-src: v-src was the first identified oncogene (Rous sarcoma virus)
	lck: Associated with the CD4 and CD8 antigens of T cells
G-protein-coupled receptors (signal transduction)	mas: Encodes the angiotensin receptor
Membrane-associated G-proteins (signal transduction)	c-ras: Three different homologues of c-ras gene, each identified in a different type of tumor and each transduced by a different retrovirus
Serine/threonine kinases (signal transduction)	c-raf: Involved in the signaling pathway; responsible for threonine phosphorylation of mitogen-activated protein (MAP) kinase following receptor activation
Nuclear DNA-binding/transcription factors	c-myc (v-myc in avian myelocytomatosis virus): Sarcomas caused by disruption of c-myc by retroviral integration or chromosomal rearrangements
	c-fos (v-fos in feline osteosarcoma virus): Interacts with a second proto-oncogene protein, Jun, to form a transcriptional regulatory complex

Table 7.2 Cell-Transforming Retroviruses

Virus Type	Time to Tumor Formation	Efficiency of Tumor Formation	Type of Oncogene
Transducing (acutely transforming)	Short (e.g., weeks)	High (up to 100%)	c-onc transduced by virus (i.e., v-onc present in virus genome; usually replication defective)
cis-Activating (chronic transforming)	Intermediate (e.g., months)	Intermediate	c-onc in cell genome activated by provirus insertion—no oncogene present in virus genome (replication competent)
trans-Activating	Long (e.g., years)	Low (<1%)	Activation of cellular genes by trans-acting virus proteins (replication competent)

normal cells. Genes involved in the formation of tumors can be grouped by their biochemical functions:

- **Oncogenes and proto-oncogenes:** Oncogenes are mutated forms of proto-oncogenes, cellular genes whose normal function is to promote the normal growth and division of cells.
- **Tumor suppressor genes:** These genes normally function to inhibit the cell cycle and cell division.

- **DNA repair genes:** These genes ensure that each strand of genetic information is accurately copied during cell division of the cell cycle. Mutations in these genes lead to an increase in the frequency of other mutations (e.g., in conditions such as ataxia–telangiectasia and xeroderma pigmentosum).

The function of oncogene products depends on their cellular location (Figure 7.4). Several classes of oncogenes are associated with the process of signal transduction—the transfer of information derived from the binding of extracellular ligands to cellular receptors to the nucleus (Figure 7.5). Many of the kinases in these groups have a common type of structure with conserved functional domains representing the hydrophobic transmembrane and hydrophilic intracellular kinase regions (Figure 7.6). These proteins are associated with the cell membranes or are present in the cytoplasm. Other classes of oncogenes located in the nucleus are normally involved with the control of the cell cycle (Figure 7.7). The products of these genes overcome the restriction between the G1 and S phases of the cell cycle, which is the key control point in preventing uncontrolled cell division. Some virus oncogenes are not sufficient on their own to produce a fully transformed phenotype in

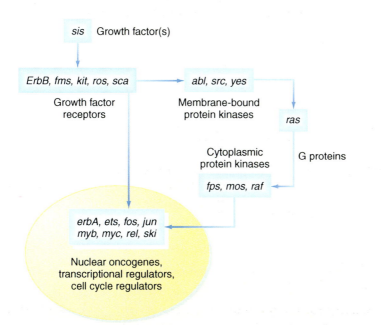

FIGURE 7.4 Subcellular location of oncoproteins.
The function of most oncogene products depends on their cellular location, for example, signal transduction, transcription factors, etc.

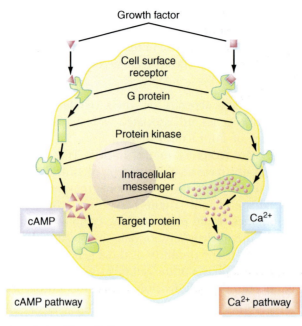

FIGURE 7.5 Cellular mechanism of signal transduction.
Several classes of oncogenes are associated with the process of signal transduction—the transfer of information derived from the binding of extracellular ligands to cellular receptors to the nucleus.

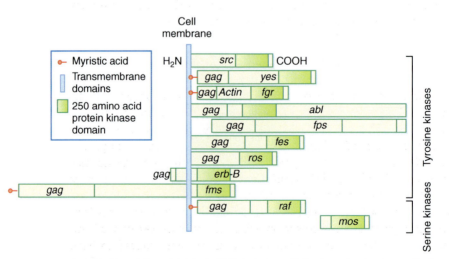

FIGURE 7.6 Retrovirus protein kinases involved in cell transformation.
Many of these molecules are fusion proteins containing amino-terminal sequences derived from the *gag* gene of the virus. Most of this type contain the fatty acid myristate which is added to the *N*-terminus of the protein after translation and which links the protein to the inner surface of the host-cell cytoplasmic membrane. In a number of cases, it has been shown that this posttranslational modification is essential to the transforming action of the protein.

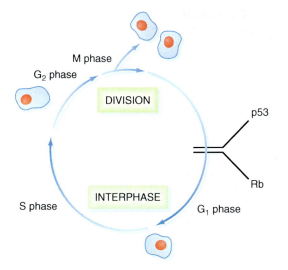

FIGURE 7.7 Phases of the eukaryotic cell cycle.
Schematic diagram showing the paths of the eukaryotic cell cycle discussed in the text.

cells; however, in some instances, they may cooperate with another oncogene of complementary function to produce a fully transformed phenotype—for example, the adenovirus E1A gene plus either the E1B gene or the c-*ras* gene transforms NIH3T3 cells (a mouse fibroblast cell line). This further underlines the fact that oncogenesis is a complex, multistep process.

CELL TRANSFORMATION BY RETROVIRUSES

Not all retroviruses are capable of transforming cells—for example, lentiviruses such as HIV do not transform cells, although they are cytopathic. The retroviruses that can transform cells fall into three groups: transducing, *cis*-activating, and *trans*-activating. The characteristics of these groups are given in Table 7.2. If oncogenes are present in all cells, why does transformation occur as a result of virus infection? The reason is that oncogenes may become activated in one of two ways, either by subtle changes to the normal structure of the gene or by interruption of the normal control of expression. The transforming genes of the acutely transforming retroviruses (v-*oncs*) are derived from and are highly homologous to c-*oncs* and are believed to have been transduced by viruses; however, most v-*oncs* possess slight alterations from their c-*onc* progenitors. Many contain minor sequence alterations that alter the structure and the function of the oncoprotein produced. Others contain short deletions of part of the gene. Most oncoproteins from replication-defective, acutely transforming retroviruses are fusion proteins, containing additional sequences derived from virus genes, most commonly virus *gag* sequences at the amino-terminus of the protein. These additional sequences may alter the function or the cellular localization of the protein, and these abnormal attributes result in transformation.

Alternatively, viruses may result in abnormal expression of an unaltered oncoprotein. This might be either the overexpression of an oncogene under the control of a virus **promoter** rather than its normal promoter in the cell, or it may be the inappropriate temporal expression of an oncoprotein that disrupts the cell cycle. Chronic transforming retrovirus **genomes** do not contain oncogenes. These viruses activate c-*oncs* by a mechanism known as insertional activation. A **provirus** that integrates into the host-cell genome close to a c-*onc* sequence may indirectly activate the expression of the gene in a way analogous to that in which v-*oncs* have been activated by transduction (Figure 7.8). This can occur if the provirus is integrated upstream of the c-*onc* gene, which might be expressed via a read-through transcript of the virus genome plus downstream sequences; however, insertional activation can also occur when a provirus integrates downstream of a c-*onc* sequence or upstream but in an inverted orientation. In these cases, activation results from **enhancer elements** in the virus promoter (see Chapter 5). These can act even if the provirus integrates at a distance of several kilobases from the c-*onc* gene. The best-known examples of this phenomenon occur in chickens, where insertion of avian leukosis virus activates the *myc* gene, and in mice, where mouse mammary tumor virus (MMTV) insertion activates the *int* gene.

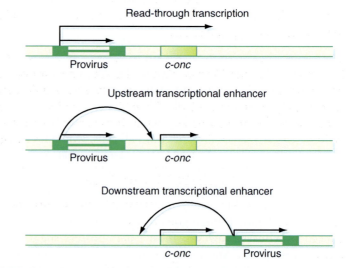

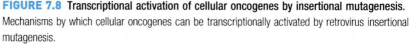

FIGURE 7.8 Transcriptional activation of cellular oncogenes by insertional mutagenesis.
Mechanisms by which cellular oncogenes can be transcriptionally activated by retrovirus insertional mutagenesis.

Transformation by the third class of retroviruses operates by quite a different mechanism. Human T-cell leukemia virus (HTLV) and related animal viruses encode a transcriptional activator protein in the virus *tax* gene. The Tax protein acts in *trans* to stimulate transcription from the virus long terminal repeat. It is believed that the protein also activates transcription of many cellular genes by interacting with transcription factors (Chapter 5); however, HTLV oncogenesis (i.e., the formation of a leukemic tumor) has a latent period of some 20−30 years. Therefore, cell transformation (which can be mimicked *in vitro*) and tumor formation (which cannot) are not one and the same—additional events are required for the development of leukemia. It is thought that chromosomal abnormalities that may occur in the population of HTLV-transformed cells are also required to produce a malignant tumor, although because of the difficulties of studying this lengthy process this is not completely understood.

CELL TRANSFORMATION BY DNA VIRUSES

In contrast to the oncogenes of retroviruses, the transforming genes of DNA tumor viruses have no cellular counterparts. Several families of DNA viruses are capable of transforming cells (Table 7.3). In general terms, the functions of their oncoproteins are much less diverse than those encoded by retroviruses. They are mostly nuclear proteins involved in the control of DNA replication which directly affect the cell cycle. They achieve their effects by interacting with cellular proteins which normally appear to have a negative regulatory role in cell proliferation. Two of the most important cellular proteins involved are known as p53 and Rb.

p53 was originally discovered by virtue of the fact that it forms complexes with SV40 T-antigen. It is now known that it also interacts with other DNA virus oncoproteins, including those of adenoviruses and papillomaviruses. The gene encoding p53 is mutated or altered in the majority of tumors, implying that loss of the normal gene product is associated with the emergence of malignantly transformed cells. Tumor cells, when injected with the

Table 7.3 Transforming Proteins of DNA Tumor Viruses

Virus	Transforming Protein(s)	Cellular Target
Adenoviruses	E1A + E1B	Rb, p53
Polyomaviruses (SV40)	T-antigen	p53, Rb
Papillomaviruses:	E5	PDGF receptor
BPV-1	E6	p53
HPV-16, 18	E7	Rb

native protein *in vitro*, show a decreased rate of cell division and decreased tumorigenicity *in vivo*. **Transgenic** mice that do not possess an intact p53 gene are developmentally normal but are susceptible to the formation of spontaneous tumors; therefore, it is clear that p53 plays a central role in controlling the cell cycle. It is believed to be a tumor suppressor or "antioncogene" and has been called the "guardian of the genome." p53 is a transcription factor that activates the expression of certain cellular genes, notably WAF1, which encodes a protein that is an inhibitor of G1 cyclin-dependent kinases, causing the cell cycle to arrest at the G1 phase (Figure 7.7). Because these viruses require ongoing cellular DNA replication for their own propagation, this explains why their transforming proteins target p53.

Rb was discovered when it was noticed that the gene that encodes this protein is always damaged or deleted in a tumor of the optic nerve known as retinoblastoma; therefore, the normal function of this gene is also thought to be that of a tumor suppressor. The Rb protein forms complexes with a transcription factor called E2F. This factor is required for the transcription of adenovirus genes, but E2F is also involved in the transcription of cellular genes which drive quiescent cells into S phase. The formation of inactive E2F–Rb complexes thus has the same overall effect as the action of p53—arrest of the cell cycle at G1. Release of E2F by replacement of E2F–Rb complexes with E1A–Rb, T-antigen–RB, or E7–RB complexes therefore stimulates cellular and virus DNA replication.

The SV40 T-antigen is one of the known virus proteins that binds p53. Chapter 5 describes the role of large T-antigen in the regulation of SV40 transcription. Infection of cells by SV40 or other polyomaviruses can result in two possible outcomes:

- Productive (**lytic**) infection
- Nonproductive (**abortive**) infection

The outcome of infection appears to be determined primarily by the cell type infected; for example, mouse polyomavirus establishes a lytic infection of mouse cells but an **abortive infection** of rat or hamster cells, while SV40 shows lytic infection of monkey cells but abortive infection of mouse cells. However, in addition to transcription, T-antigen is also involved in **genome** replication. SV40 DNA replication is initiated by binding of large T-antigen to the origin region of the genome (Figure 5.12). The function of T-antigen is controlled by phosphorylation, which decreases the ability of the protein to bind to the SV40 origin.

The SV40 **genome** is very small and does not encode all the information necessary for DNA replication; therefore, it is essential for the host cell to enter

S phase, when cell DNA and the virus genome are replicated together. Protein–protein interactions between T-antigen and DNA polymerase α directly stimulate replication of the virus genome. The precise regions of the T-antigen involved in binding to DNA, DNA polymerase α, p53, and Rb are all known (Figure 7.9). Inactivation of tumor suppressor proteins bound to T-antigen causes G1-arrested cells to enter S phase and divide, and this is the mechanism that results in transformation; however, the frequency with which abortively infected cells are transformed is low (about 1×10^{-5}). Therefore, the function of T-antigen is to alter the cellular environment to permit virus DNA replication. Transformation is a rare and accidental consequence of the sequestration of tumor suppressor proteins.

The immediate-early proteins of adenoviruses are analogous in many ways to SV40 T-antigen. E1A is a *trans*-acting transcriptional regulator of the adenovirus early genes (see Chapter 5). Like T-antigen, the E1A protein binds to Rb, inactivating the regulatory effect of this protein, permitting virus DNA replication, and accidentally stimulating cellular DNA replication. E1B binds p53 and reinforces the effects of E1A. The combined effect of the two proteins can be seen in the phenotype of cells transfected with DNA containing these genes (Table 7.4). However, the interaction of these transforming proteins with the cell is more complex than simple induction of DNA synthesis. Expression of E1A alone causes cells to undergo apoptosis. Expression of E1A and E1B together overcomes this response and permits transformed cells to survive and grow.

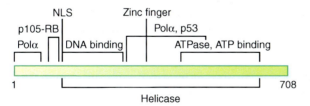

FIGURE 7.9 Regions of SV40 T-antigen involved in protein–protein interactions. Other functional domains of the protein involved in virus DNA replication are also shown, including the helicase, ATPase, and nuclear location signal (NLS) domains.

Table 7.4 Role of the Adenovirus E1A and E1B Proteins in Cell Transformation

Protein	Cell Phenotype
E1A	Immortalized but morphologically unaltered; not tumorigenic in animals
E1B	Not transformed
E1A + E1B	Immortalized and morphologically altered; tumorigenic in animals

Human papillomavirus (HPV) genital infections are very common, occurring in more than 50% of young, sexually active adults, and are usually asymptomatic. Certain serotypes of HPV appear to be associated with a low risk of subsequent development of anogenital cancers such as cervical carcinoma, after an incubation period of several decades. 500,000 new cases of cervical neoplasia are diagnosed every year, making this one of the three most common causes of cancer death in women globally. HPV is a primary cause of cervical cancer; 93% of all cervical cancers test positive for one or more high-risk type of HPV. Of the 60 HPV types currently recognized, only four seem to be associated with a high risk of tumor formation (HPV-16, 18, 31, and 45). Once again, transformation is mediated by the early gene products of the virus. However, the transforming proteins appear to vary from one type of papillomavirus to another, as shown in Table 7.3. In general terms, it appears that two or more early proteins often cooperate to give a transformed phenotype. Although some papillomaviruses can transform cells on their own (e.g., bovine papillomavirus 1, BPV-1), others appear to require the cooperation of an activated cellular oncogene (e.g., HPV-16/*ras*). In bovine papillomavirus, it is the E5 protein that is responsible for transformation. In HPV-16 and HPV-18, the E6 and E7 proteins are involved.

More confusingly, in most cases all or part of the papillomavirus genome, including the putative transforming genes, are maintained in the tumor cells, whereas in some cases (e.g., BPV-4) the virus DNA may be lost after transformation, which may indicate a possible hit-and-run mechanism of transformation. Different papillomaviruses appear to use slightly different mechanisms to achieve genome replication, so cell transformation may proceed via a slightly different route. It is imperative that a better understanding of these processes is obtained. There is no positive evidence that adenoviruses or polyomaviruses are involved in the formation of human tumors. In contrast, the evidence that papillomaviruses are commonly involved in the formation of malignant penile and cervical carcinomas is now very strong.

In recent years, evidence has emerged that p53 and Rb are major cellular sensors for apoptosis. Loss of these protein functions triggers apoptosis—the major anticancer mechanism in cells; thus, viruses that interfere with these proteins must have evolved mechanisms to counteract this effect (see discussion in Chapter 6).

VIRUSES AND CANCER

There are numerous examples of viruses that cause tumors in experimental animals, stimulating a long search for viruses that might be the cause of cancer in humans. For many years, this search was unsuccessful, so much so that a few scientists categorically stated that viruses did not cause human tumors. Like all rash statements, this one was wrong. An estimated 20% of all human cancers worldwide may be caused by viruses. Although it is convenient to consider human tumor viruses as a discrete group of viruses, the six viruses which cause human cancers have very different genomes, replication cycles, and come from five different virus families (HHV-4/EBV, HBV, HCV, HHV-8, HPVs, HTLV). The path from virus infection to tumor formation is slow and inefficient. Only a minority of infected individuals progress to cancer, usually years or even decades after primary infection. Virus infection alone is generally not sufficient for cancer, and additional events and host factors, such as immunosuppression, somatic mutations, genetic predisposition, and exposure to carcinogens, must also play a role.

The role of the HTLV tax protein in leukemia has already been described (see "Cell Transformation by Retroviruses"). The evidence that papillomaviruses may be involved in human tumors is now well established. There are almost certainly many more viruses that cause human tumors, but the remainder of this chapter describes two examples that have been intensively studied: EBV and hepatitis B virus (HBV).

EBV was first identified in 1964 in a lymphoblastoid cell line derived from an African patient with Burkitt's lymphoma. In 1962, Dennis Burkitt described a highly malignant lymphoma, the distribution of which in Africa paralleled that of malaria. Burkitt recognized that this tumor was rare in India but occurred in Indian children living in Africa and therefore looked for an environmental cause. Initially, he thought that the tumor might be caused by a virus spread by mosquitoes (which is wrong). The association between EBV and Burkitt's lymphoma is not entirely clear cut:

- EBV is widely distributed worldwide but Burkitt's lymphoma is rare.
- EBV is found in many cell types in Burkitt's lymphoma patients, not just in the tumor cells.
- Rare cases of EBV-negative Burkitt's lymphoma are sometimes seen in countries where malaria is not present, suggesting there may be more than one route to this tumor.

EBV has a dual cell tropism for human B-lymphocytes (generally a nonproductive infection) and epithelial cells, in which a productive infection occurs. The usual outcome of EBV infection is polyclonal B-cell activation and a benign proliferation of these cells which is frequently asymptomatic but sometimes produces a relatively mild disease known as infectious mononucleosis or glandular fever. In 1968, it was shown that EBV could efficiently transform (i.e., immortalize) human B-lymphocytes *in vitro*. This observation clearly strengthens the case that EBV is involved in the formation of tumors.

There is now epidemiological and/or molecular evidence that EBV infection is associated with at least five human tumors:

- Burkitt's lymphoma.
- Nasopharyngeal carcinoma (NPC), a highly malignant tumor seen most frequently in China. There is a strong association between EBV and NPC. Unlike Burkitt's lymphoma, the virus has been found in all the tumors that have been studied. Environmental factors, such as the consumption of nitrosamines in salted fish, are also believed to be involved in the formation of NPC (cf. the role of malaria in the formation of Burkitt's lymphoma).
- B-cell lymphomas in immunosuppressed individuals (e.g., AIDS patients).
- Some clonal forms of Hodgkin's disease.
- X-linked lymphoproliferative syndrome (XLP), a rare condition usually seen in males where infection with EBV results in a hyperimmune response, sometimes causing a fatal form of glandular fever and sometimes cancer of the lymph nodes. XLP is an inherited defect due to a faulty gene on the X chromosome.

Cellular transformation by EBV is a complex process involving the cooperative interactions between several viral proteins. Three possible explanations for the link between EBV and Burkitt's lymphoma are:

1. EBV immortalizes a large pool of B-lymphocytes; concurrently, malaria causes T-cell immunosuppression. There is thus a large pool of target cells in which a third event (e.g., a chromosomal translocation) results in the formation of a malignantly transformed cell. Most Burkitt's lymphoma tumors contain translocations involving chromosome 8, resulting in activation of the c-*myc* gene, which supports this hypothesis.
2. Malaria results in polyclonal B-cell activation. EBV subsequently immortalizes a cell containing a preexisting c-*myc* translocation. This mechanism would be largely indistinguishable from the above.
3. EBV is just a passenger virus! Burkitt's lymphoma also occurs in Europe and North America although it is very rare in these regions; however, 85% of these patients are not infected with EBV, which implies that there are other causes for Burkitt's lymphoma.

Although it has not been formally proved, it seems likely that either (1) and/or (2) is the true explanation for the origin of Burkitt's lymphoma.

Another case where a virus appears to be associated with the formation of a human tumor is that of HBV and hepatocellular carcinoma (HCC). Hepatitis is an inflammation of the liver and as such is not a single disease. Because of the central role of the liver in metabolism, many virus infections may involve the liver; however, at least seven viruses seem specifically to infect and damage hepatocytes. No two of these belong to the same family (see Chapter 8). HBV is the prototype member of the family *Hepadnaviridae* and causes the disease formerly known as "serum hepatitis." This disease was distinguished clinically from "infectious hepatitis" (caused by other types of hepatitis virus) in the 1930s. HBV infection formerly was the result of inoculation with human serum (e.g., blood transfusions, organ transplants) but is still common among intravenous drug abusers, where it is spread by the sharing of needles and syringes; however, the virus is also transmitted sexually, by oral ingestion, and from mother to child, which accounts for familial clusters of HBV infection. All blood, organ, and tissue donations in developed countries are now tested for HBV, and risk of transmission is extremely low. The virus does not replicate in tissue culture which has seriously hindered investigations into its pathogenesis. HBV infection has three possible outcomes:

1. An acute infection followed by complete recovery and immunity from reinfection (>90% of cases).
2. Fulminant hepatitis, developing quickly and lasting a short time, causing liver failure and a mortality rate of approximately 90% (<1% of cases).
3. Chronic infection, leading to the establishment of a carrier state with virus persistence (about 10% of cases).

There are approximately 350 million chronic HBV carriers worldwide. The total population of the world is approximately 6 billion; therefore, about 5% of the world population is persistently infected with HBV. All of these chronic carriers of the virus are at 100−200 times the risk of noncarriers of developing HCC. HCC is a rare tumor in the West, where it represents <2% of fatal cancers. Most cases that do occur in the West are alcohol related, and this is an important clue to the pathogenesis of the tumor; however, in Southeast Asia and in China, HCC is the most common fatal cancer, resulting in about half a million deaths every year. The virus might cause the formation of the tumor by three different pathways: direct activation of a cellular oncogene(s), *trans*-activation of a cellular oncogene(s), or indirectly via tissue regeneration (Figure 7.10). As with EBV and Burkitt's lymphoma, the relationship between HBV and HCC is not clear cut:

- Cirrhosis (a hardening of the liver which may be the result of infections or various toxins, such as alcohol) appears to be a prerequisite for the development of HCC. It would appear that chronic liver damage induces tissue regeneration and that faulty DNA repair mechanisms result eventually in malignant cell transformation. Unrelated viruses

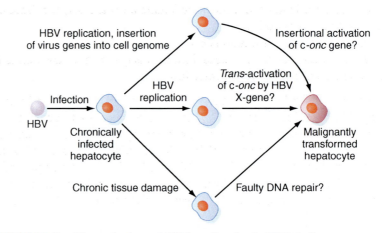

FIGURE 7.10 **Possible mechanisms of HCC formation due to HBV infection.**
The complex relationship between HBV infection and HCC means that it is not certain whether any or all of these possible mechanisms are involved.

that cause chronic active hepatitis, such as the flavivirus hepatitis C virus (HCV), are also associated with HCC after a long latent period.

- A number of cofactors, such as aflatoxins and nitrosamines, can induce HCC-like tumors in experimental animals without virus infection; therefore, such substances may also be involved in human HCC (cf. nitrosamines and NPC, above).

For many years, it was thought that HBV integration events were random with regard to their sites within the human genome, but when the relationship between "fragile sites" in the host genome and virus integration events are compared, HBV DNA is found to integrate within or near many of these fragile regions. In most cases, integration at a particular site has only been reported for a single or small number of tumors, but a closer look shows that individual integration sites alters the expression of different components in the same or redundant biochemical or signaling pathways that support hepatocellular growth and survival important for tumor development. Most (but not all) HBV integration events retain the open reading frame encoding the HBx antigen (HBxAg), which suggests that this protein contributes to HCC, possibly via miRNAs. It is possible that all the mechanisms shown in Figure 7.10 might operate *in vivo*. The key risk factor is the development of a chronic as opposed to an acute HBV infection. This in itself is determined by a number of other factors:

- Age—The frequency of chronic infections declines with increasing age at the time of infection.

- Sex—For chronic infection, the male:female ratio is 1.5:1; for cirrhosis, the male:female ratio is 3:1.
- HCC—The male:female ratio is 6:1.
- Route of infection—Oral or sexual infections give rise to fewer cases of chronic infection than serum infection.

Until there is a much better understanding of the pathogenesis and normal course of HBV infection, it is unlikely that the reasons for these differences will be understood. There may be a happy ending to this story. A safe and effective vaccine that prevents HBV infection is now available and widely used in the areas of the world where HBV infection is endemic as part of the WHO Expanded Programme on Immunization. This will prevent a million deaths annually from HCC and HBV disease in the future.

NEW AND EMERGENT VIRUSES

What constitutes a "new" infectious agent? Are these just viruses that have never been discovered, or are they previously known viruses that have changed their behavior? This section will describe and attempt to explain current understanding of a number of agents that meet the above criteria. In the century, massive and unexpected epidemics have been caused by certain viruses. For the most part, these epidemics have not been caused by completely new (i.e., previously unknown) viruses but by viruses that were well known in the geographical areas in which they may currently be causing epidemic outbreaks of disease. Such viruses are known as emergent viruses (Table 7.5). There are numerous examples of such viruses that appear to have mysteriously altered their behavior with time, with significant effects on their pathogenesis.

One of the better known examples of this phenomenon is poliovirus. It is known that poliovirus and poliomyelitis have existed in human populations for at least 4,000 years. For most of this time, the pattern of disease was endemic rather than epidemic (i.e., a low, continuous level of infection in particular geographical areas. During the first half of the twentieth century, the pattern of occurrence of poliomyelitis in Europe, North America, and Australia changed to an epidemic one, with vast annual outbreaks of infantile paralysis. Although we do not have samples of polioviruses from earlier centuries, the clinical symptoms of the disease give no reason to believe that the virus changed substantially. Why, then, did the pattern of disease change so dramatically? It is believed that the reason is as follows. In rural communities with primitive sanitation facilities, poliovirus circulated freely. Serological surveys in similar contemporary situations reveal that more than 90% of children of 3 years of age have antibodies to at least one of the three

Table 7.5 Some Examples of Emergent Viruses

Virus	Family	Comments
Cocoa swollen shoot	Badnavirus	Emerged in 1936 and is now the main disease of cocoa in Africa. Deforestation increases population of mealy bug vectors and disease transmission.
Hendra virus	Paramyxovirus	Emerged in Brisbane, Australia, September 1994. Causes acute respiratory disease in horses with high mortality and a fatal encephalitis in humans—with several deaths so far. The disease, normally carried by fruit bats (with no pathogenesis), has reemerged in humans in Queensland several times since 1994.
Nipah virus	Paramyxovirus	Emerged in Malaysia in 1998. Closely related to Hendra virus; a zoonotic virus transmitted from animals (pigs?) to humans. Mortality rate in outbreaks of up to 70%.
Phocine distemper	Paramyxovirus	Emerged in 1987 and caused high moralities in seals in the Baltic and North Seas. Similar viruses subsequently recognized as responsible for cetacean (porpoise and dolphin) deaths in Irish Sea and Mediterranean. The virus was believed to have been introduced into immunologically naive seal populations by a massive migration of harp seals from the Barents Sea to northern Europe.
Rabbit hemorrhagic disease (RHD), also known as rabbit calicivirus disease (RCD) or viral hemorrhagic disease (VHD)	Calicivirus	Emerged in farmed rabbits in China in 1984, spread through the United Kingdom, Europe, and Mexico. Introduced to Wardang Island off the coast of South Australia to test potential for rabbit population control, the disease accidentally spread to Australian mainland, causing huge kill in rabbit populations. A vaccine is available to protect domestic and farmed rabbits. In August 1997, RHD was illegally introduced into the South Island of New Zealand and escaped into the United States in April 2000.

serotypes of poliovirus. (Even the most virulent strains of poliovirus cause 100–200 subclinical infections for each case of the paralytic poliomyelitis seen.) In such communities, infants experience subclinical immunizing infections while still protected by maternal antibodies—a form of natural vaccination. The relatively few cases of paralysis and death that do occur are likely to be overlooked, especially in view of high infant mortality rates. During the nineteenth century, industrialization and urbanization changed the pattern of poliovirus transmission. Dense urban populations and increased traveling afforded opportunities for rapid transmission of the virus. In addition, improved sanitation broke the natural pattern of virus transmission. Children were likely to encounter the virus for the first time at a later age and without the protection of maternal antibodies. These children were at far greater risk when they did eventually become infected, and it is believed that these social changes account for the altered pattern of disease. Fortunately, the widespread use of poliovirus vaccines has since brought the situation under control in industrialized countries (Chapter 6). In 1988, the World Health Organization committed itself to wiping out polio completely ("eradication") by the year 2000. But the disease has proved to be troublingly resilient in a few of the poorest, more corrupt and most dangerous countries, and is still hanging on. Polio eradication is no longer a technical challenge, rather it is a political and economic one.

There are many examples of the epidemic spread of viruses caused by movement of human populations. Measles and smallpox were not known to the ancient Greeks. Both of these viruses are maintained by direct person-to-person transmission and have no known alternative hosts; therefore, it has been suggested that it was not until human populations in China and the Roman Empire reached a critical density that these viruses were able to propagate in an epidemic pattern and cause recognizable outbreaks of disease. Before this time, the few cases that did occur could easily have been overlooked. Smallpox reached Europe from the Far East in 710 AD, and in the eighteenth century it achieved plague proportions—five reigning European monarchs died from smallpox. However, the worst effects occurred when these viruses were transmitted to the New World. Smallpox was (accidentally) transferred to the Americas by Hernando Cortés in 1520. In the next two years, 3.5 million Aztecs died from the disease and the Aztec empire was decimated by disease rather than conquest. Although not as highly pathogenic as smallpox, epidemics of measles subsequently finished off the Aztec and Inca civilizations. More recently, the first contacts with isolated groups of Eskimos and tribes in New Guinea and South America have had similarly devastating results, although on a smaller scale. These historical incidents illustrate the way in which a known virus can suddenly cause illness and death on a catastrophic scale following a change in human behavior.

Measles and smallpox viruses are transmitted exclusively from one human host to another. For viruses with more complex cycles of transmission (e.g., those with secondary hosts and insect vectors), control of infection becomes much more difficult (Figure 7.11). This is particularly true of the families of viruses known collectively as "arboviruses" (arenaviruses, bunyaviruses, flaviviruses, and togaviruses). As human territory has expanded, this has increasingly brought people into contact with the type of environment where these viruses are found—warm, humid, vegetated areas where insect vectors occur in high densities, such as swamps and jungles.

A classic example is the mortality caused by yellow fever virus during the building of the Panama Canal at the end of the nineteenth century. More recently, the increasing pace of ecological alteration in tropical areas has resulted in the resurgence of yellow fever in Central America, particularly an

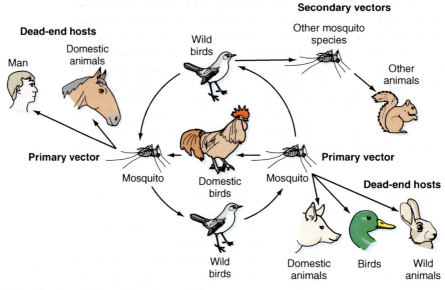

FIGURE 7.11 Complex transmission pattern of an arbovirus.

Because of their complex transmission patterns involving multiple host species, arthropod-borne viruses are difficult to control, let alone to eradicate.

urban form of the disease transmitted directly from one human to another by mosquitoes. Dengue fever is also primarily an urban disease of the tropics, transmitted by *Aedes aegypti*, a domestic, day-biting mosquito that prefers to feed on humans. Some outbreaks of dengue fever have involved more than a million cases, with attack rates of up to 90% of the population. There are believed to be over 40 million cases of dengue virus infection worldwide each year. This disease was first described in 1780. By 1906, it was known that the virus was transmitted by mosquitoes, and the virus was isolated in 1944; therefore, this is not a new virus, but the frequency of dengue virus infection has increased dramatically in the last 30 years due to changes in human activity.

Of more than 500 arboviruses known, at least 100 are pathogenic for humans and at least 20 would meet the criteria for emergent viruses. Attempts to control these diseases rely on twin approaches involving both the control of insect vectors responsible for transmission of the virus to humans and the development of vaccines to protect human populations. However, both of these approaches present considerable difficulties, the former in terms of avoiding environmental damage and the latter in terms of understanding virus pathogenesis and developing appropriate vaccines (see earlier discussion of dengue virus pathogenesis). Rift Valley fever virus (RVFV) was first isolated from sheep in 1930 but has caused repeated epidemics in sub-Saharan Africa during the last few decades, with human infection rates in epidemic areas as high as 35%. This is an epizootic disease, transmitted from sheep to humans by a number of different mosquitoes. The construction of dams which increase mosquito populations, increasing numbers of sheep, and the movement of sheep and human populations are believed to be responsible for the upsurge in this disease. RVFV continues to extend its range in Africa and the Middle East and is a significant health and economic burden in many areas of Africa, remaining a serious threat to other parts of the world.

The *Hantavirus* genus (*Bunyaviridae*) is a particular cause for concern. Hantaviruses cause millions of cases of hemorrhagic fever each year in many parts of the world. Unlike arboviruses, hantaviruses are transmitted directly from rodent hosts to humans (e.g., via feces) rather than by an invertebrate host. Hantaviruses cause two acute diseases: hemorrhagic fever with renal syndrome (HFRS) and hantavirus pulmonary syndrome (HPS). HFRS was first recognized in 1951 after an outbreak among U.S. troops stationed in Korea. In 1993, HPS was first recognized in the United States, and a new virus, Sin Nombre, was identified as the cause. It is now known that at least three different hantaviruses cause HFRS and four different viruses cause HPS. By 1995, HPS had been recognized in 102 patients in 21 states of the United States, in seven patients in Canada, and in three in Brazil, with an overall mortality rate of approximately 40%. These statistics illustrate the disease-causing potential of emerging viruses.

West Nile virus (WNV) is a member of the Japanese encephalitis antigenic complex of the family *Flaviviridae*. All known members of this complex are transmissible by mosquitoes, and many of them can cause febrile, sometimes fatal, illnesses in humans. WNV was first isolated in the West Nile district of Uganda in 1937 but is in fact the most widespread of the flaviviruses, with geographic distribution including Africa and Eurasia. Unexpectedly, an outbreak of human encephalitis caused by WNV occurred in the United States in New York and surrounding states in 1999. In this case, the virus appears to have been transmitted from wild, domestic, and exotic birds by *Culex* mosquitoes (an urban mosquito that flourishes under dry conditions)—a classic pattern of arbovirus transmission. WNV RNA has been detected in overwintering mosquitoes and in birds, and the disease is now endemic across the United States, causing outbreaks each summer. This rapid spread into a new territory shows that spread did not rely on environmental factors such as climate change—the North American environment was already suitable for the virus once it had been introduced, probably via air travel from the Middle East.

Chikungunya virus (CHIKV) is transmitted by *Aedes* mosquitoes and was first isolated in 1953 in Tanzania. CHIKV is a member of the genus *Alphavirus* and the family *Togaviridae*. The disease caused by this virus typically consists of an acute illness characterized by fever, rash, and incapacitating joint pain. The word chikungunya means "to walk bent over" in some east African languages and refers to the effect of the joint pains that characterize this dengue-like infection. Chikungunya is a specifically tropical disease, but was previously geographically restricted and outbreaks were relatively uncommon. The virus remained largely unknown until a major outbreak in 2005 and 2006 on islands across the Indian Ocean. Plausible explanations for this outbreak (and subsequent spread, which has continued) include increased tourism, CHIKV introduction into a naive population, and virus mutation. It is the last of those three factors which seems to be most significant in this case, with the outbreak strain showing a single amino acid change in the envelope glycoprotein which allows more effective transmission due to more efficient crossing of the mosquito gut membrane barrier. There is every possibility that CHIKV will continue to extend its territory.

Plant viruses can also be responsible for **emergent diseases**. Group III geminiviruses are transmitted by insect vectors (whiteflies), and their **genomes** consist of two circular, single-stranded DNA molecules (Chapter 3). These viruses cause a great deal of crop damage in plants such as tomatoes, beans, squash, cassava, and cotton, and their spread may be directly linked to the inadvertent worldwide dissemination of a particular biotype of the whitefly *Bemisia tabaci*. This vector is an indiscriminate feeder, encouraging the rapid and efficient spread of viruses from indigenous plant species to neighboring crops.

Occasionally, there appears an example of an emergent virus that has acquired extra genes and as a result of this new genetic capacity has become capable of infecting new species. A possible example of this phenomenon is seen in tomato spotted wilt virus (TSWV). TSWV is a bunyavirus with a very wide plant host range, infecting over 600 different species from 70 families. In recent decades, this virus has been a major agricultural pest in Asia, the Americas, Europe, and Africa. Its rapid spread has been the result of dissemination of its insect vector (the thrip *Frankinellia occidentalis*) and diseased plant material. TSWV is the type species of the *Tospovirus* genus and has a morphology and genomic organization similar to the other bunyaviruses (Chapter 3). However, TSWV undergoes **propagative transmission**, and it has been suggested that it may have acquired an extra gene in the M segment via **recombination**, either from a plant or from another plant virus. This new gene encodes a **movement protein** (Chapter 6), conferring the capacity to infect plants and cause extensive damage.

In addition to viruses whose ability to infect their host species appears to have changed, new viruses are being discovered continually. After many years of study, three new HHV have been discovered comparatively recently:

- **HHV-6:** First isolated in 1986 in lymphocytes of patients with lymphoreticular disorders; tropism for CD4$^+$ lymphocytes. HHV-6 is now recognized as being an almost universal human infection. Discovery of the virus solved a longstanding mystery: The primary infection in childhood causes "roseola infantum" or "fourth disease," a common childhood rash of previously unknown cause. Antibody titers are highest in children and decline with age. The consequences of childhood infection appear to be mild. Primary infections of adults are rare but have more severe consequences—mononucleosis or hepatitis— and infections may be a severe problem in transplant patients.
- **HHV-7:** First isolated from human CD4$^+$ cells in 1990. Its genome organization is similar to but distinct from that of HHV-6, and there is limited antigenic cross-reactivity between the two viruses. Currently, there is no clear evidence for the direct involvement of HHV-7 in any human disease, but might it be a cofactor in HHV-6-related syndromes.
- **HHV-8:** In 1995, sequences of a unique herpesvirus were identified in DNA samples from AIDS patients with KS and in some non-KS tissue samples from AIDS patients. There is a strong correlation (>95%) with KS in both HIV$^+$ and HIV2 patients. HHV-8 can be isolated from lymphocytes and from tumor tissue and appears to have a less ubiquitous world distribution than other HHVs; that is, it may only be associated with a specific disease state (cf. HSV, EBV). However, the virus is not present in KS-derived cell lines, suggesting that autocrine or paracrine factors may be involved in the formation of KS. There is some evidence that HHV-8 may also cause other tumors such as B-cell lymphomas ($\pm$ EBV as a "helper").

Although many different virus infections may involve the liver, at least six viruses seem specifically to infect and damage hepatocytes. No two of these belong to the same family! The identification of these viruses has been a long story:

- Hepatitis B virus (HBV) (hepadnavirus): 1963
- Hepatitis A virus (HAV) (picornavirus): 1973
- Hepatitis delta virus (HDV) (deltavirus; see Chapter 8): 1977
- Hepatitis C virus (HCV) (flavivirus): 1989
- Hepatitis E virus (HEV): 1990
- GBV-C/HGV: 1995
- "Transfusion-transmitted virus" (TTV): 1998

Reports continue to circulate about the existence of other hepatitis viruses. Some of the agents are reported to be sensitive to chloroform (i.e., **enveloped**) while others are not. This may suggest the existence of multiple viruses, as yet undescribed, although this is still uncertain. Although the viral causes of the majority of cases of infectious human hepatitis have now been identified, it is possible that further hepatitis viruses will be described in the future. New human retroviruses are being discovered regularly, some of them of great significance:

- HTLV: 1981
- HIV: 1983
- XMRV: 2006

BOX 7.3 WHERE DO VIRUSES COME FROM?

In spite of what a few people believe, there's no evidence they come from outer space (strike one for the alien abduction theory of virology). So either they come from preexisting viruses which change in some way, or they were there all the time and we just didn't notice them. That's not as stupid as it sounds. Using molecular clock built into virus genomes researchers have been able to show pretty convincingly that viruses such as measles seemed to pop up just at the point when human populations were big enough to support them by continuous person-to-person spread. And so a cow virus (rinderpest) became a human virus (measles). Like smallpox before them, both measles and rinderpest are now on the verge of complete eradication. But don't get too excited. Just as monkeypox seems to be evolving into the old niche that smallpox filled in Africa, there'll be another virus along to replace measles pretty soon.

ZOONOSES

Many **emergent virus diseases** are **zoonoses** (i.e., transmitted from animals to humans). This emphasizes the importance of the "species barrier" in preventing transmission of infectious diseases, and several recent examples illustrate the potentially disastrous consequences that can occur when this is breached. Strictly speaking, many of the "arboviruses" discussed earlier are zoonotic in humans, but their transmission involves an insect vector. On occasions, viruses can spread from animals into the human population and then be transmitted from one person to another without the involvement of a vector.

Severe acute respiratory syndrome (SARS) is a type of viral pneumonia, with symptoms including fever, a dry cough, shortness of breath, and headaches. Death may result from progressive respiratory failure due to lung damage. The first SARS outbreak originated in the Guangdong province of China in 2003, where 300 people became ill and at least five died. The cause was found to be a novel coronavirus, SARS-CoV. The SARS virus is believed to be spread by droplets produced by coughing and sneezing, but other routes of infection may also be involved, such as fecal contamination. Where did the SARS virus come from? Coronaviruses with 99% sequence similarity to the surface spike protein of human SARS isolates have been isolated in Guangdong, China, from apparently healthy masked palm civets, a cat-like mammal closely related to the mongoose. The unlucky palm civet is regarded as a delicacy in Guangdong and it is believed that humans became infected as they raised and slaughtered the animals rather than by consumption of infected meat. After 2003, SARS went away, but in 2012, a related coronavirus, Middle East respiratory syndrome coronavirus (MERS-CoV), popped up in the Middle East. Like SARS, MERS is a zoonotic infection, probably originating in bats but also found in camels. It is not clear whether the emergence of MERS is the result of a single zoonotic event with subsequent human-to-human transmission, or if the multiple geographic sites of infection across the Middle East represent multiple zoonotic events from a common source. MERS-CoV is less infectious than SARS-CoV but has still infected nearly a thousand people and killed 350 to date.

Ebola virus was first identified in 1976. Isolates from Central Africa appear to be highly pathogenic, whereas those from the Philippines are less pathogenic for humans. The molecular basis for these differences is unknown. Most Ebola virus outbreaks appear to be associated with contact with infected primates; however, extensive ecological surveys in Central Africa have failed to show any evidence that primates (or any of the thousands of animals, plants, and invertebrate species examined) are the natural reservoir for infection. No animal reservoir for the virus has been positively identified, but fruit and insectivorous bats support replication and circulation of high titers of Ebola virus without necessarily becoming ill. As with SARS, consumption of exotic wild meats (called "bushmeat"), particularly primates, may be a risk factor. The most recent outbreak in West Africa which began in 2013 and is still continuing is by far the largest to date, having killed over 9,000 people, but fortunately not the most virulent as the death rate is around 60%, compared with over 90% in some smaller previous outbreaks.

Other new zoonotic viruses are frequently discovered, fortunately rarely with the serious disease potential of SARS or Ebola virus.

BIOTERRORISM

Along with the threats from emerging viruses, the world currently faces the potential use of viruses as terrorist weapons. Although this issue has received much media attention, the reality is that the deliberate releases of such pathogens may have less medical impact than is generally appreciated. Many governments devoted considerable resources to the development of viruses as weapons of war before deciding that their military usefulness was very limited. The U.S. Centers for Disease Control (CDC) only recognizes two types of virus as potentially dangerous terrorist weapons: smallpox and agents causing hemorrhagic fevers such as filoviruses and arenaviruses. Emerging viruses such as Nipah virus and hantaviruses are also recognized as possible future threats. However, this is in contrast to a much larger number of bacterial species and toxins. The reason for this is that bacterial pathogens would be much easier for terrorist groups to prepare and disseminate than viruses. The potential threat from bioterrorism is in reality insignificant in relation to the actual number of deaths caused by infections worldwide each year. Nevertheless, this is an issue which governments are sensibly treating with great seriousness.

SUMMARY

Virus pathogenesis is a complex, variable, and relatively rare state. Like the course of a virus infection, pathogenesis is determined by the balance between host and virus factors. Not all of the pathogenic symptoms seen in virus infections are caused directly by the virus—the immune system also plays a part in causing cell and tissue damage. Viruses can transform cells so that they continue to grow indefinitely. In some but not all cases, this can lead to the formation of tumors. There are some well-established cases where certain viruses provoke human tumors and possibly many others that we do not yet understand. The relationship between the virus and the formation of the tumor is not a simple one, but the prevention of infection undoubtedly reduces the risk of tumor formation. New pathogenic viruses are being discovered all the time, and changes in human activities result in the emergence of new or previously unrecognized diseases.

Further Reading

Angel, R.M., Valle, J.R., 2013. Dengue vaccines: strongly sought but not a reality just yet. PLoS Pathog. 9 (10), e1003551. Available from: http://dx.doi.org/10.1371/journal.ppat.1003551.

Archin, N.M., Sung, J.M., Garrido, C., Soriano-Sarabia, N., Margolis, D.M., 2014. Eradicating HIV-1 infection: seeking to clear a persistent pathogen. Nat. Rev. Microbiol. 12 (11), 750−764.

Best, S.M., 2008. Viral subversion of apoptotic enzymes: escape from death row. Annu. Rev. Microbiol. 62, 171−192.

Casadevall, A., Pirofski, L.A., 2004. The weapon potential of a microbe. Trends Microbiol. 12, 259−263.

Cubie, H.A., 2013. Diseases associated with human papillomavirus infection. Virology 445 (1), 21−34.

DeCaprio, J.A., 2009. How the Rb tumor suppressor structure and function was revealed by the study of adenovirus and SV40. Virology 384 (2), 274−284.

Donlan, R.M., 2009. Preventing biofilms of clinically relevant organisms using bacteriophage. Trends Microbiol. 17 (2), 66−72.

Forsman, A., Weiss, R.A., 2008. Why is HIV a pathogen? Trends Microbiol. 16 (12), 555−560.

Goeijenbier, M., van Kampen, J.J., Reusken, C.B., Koopmans, M.P.G., van Gorp, E.C.M., 2014. Ebola virus disease: a review on epidemiology, symptoms, treatment and pathogenesis. Neth. J. Med. 72 (9), 242−248.

LaBeaud, A.D., Kazura, J.W., King, C.H., 2010. Advances in Rift Valley fever research: insights for disease prevention. Curr. Opin. Infect. Dis. 23 (5), 403−408.

Morales-Sánchez, A., Fuentes-Pananá, E.M., 2014. Human viruses and cancer. Viruses 6 (10), 4047−4079.

Nash, A.A., Dalziel, R.G., Fitzgerald, J.R., 2015. Mims' Pathogenesis of Infectious Disease. Academic Press, ISBN 0123971888.

Racaniello, V.R., 2006. One hundred years of poliovirus pathogenesis. Virology 344 (1), 9−16.

Randolph, S.E., Rogers, D.J., 2010. The arrival, establishment and spread of exotic diseases: patterns and predictions. Nat. Rev. Microbiol. 8 (5), 361−371.

Thorley-Lawson, D.A., Allday, M.J., 2008. The curious case of the tumour virus: 50 years of Burkitt's lymphoma. Nat. Rev. Microbiol. 6 (12), 913−924.

Weaver, S.C., Reisen, W.K., 2010. Present and future arboviral threats. Antiviral Res. 85 (2), 328−345.

Whitehorn, J., Simmons, C.P., 2011. The pathogenesis of dengue. Vaccine 29 (42), 7221−7228.

CHAPTER 8

Subviral Agents: Genomes without Viruses, Viruses without Genomes

Intended Learning Outcomes

On completing this chapter you should be able to:

- Discuss the minimal genome needed by a living entity.
- Explain what satellites and viroids are and how they differ from viruses.
- Describe how prions, infectious protein molecules seemingly with no genome at all, can cause disease.

What is the minimum **genome** size necessary to sustain an infectious agent? Could a virus with a genome of 1,700 nt survive? Or a genome of 240 nt? Could an infectious agent without any genome at all exist? Perhaps the first two alternatives might be possible, but the idea of an infectious agent without a genome seems bizarre and ridiculous. Strange as it may seem, such agents as these do exist and cause disease in animals (including humans) and plants.

SATELLITES AND VIROIDS

Satellites are small RNA molecules that are absolutely dependent on the presence of another virus for multiplication. Even viruses have their own parasites! Most satellites are associated with plant viruses, but a few are associated with **bacteriophages** or animal viruses (e.g., the *Dependovirus* genus) that are satellites of adenoviruses. Two classes of satellites can be distinguished: satellite viruses, which encode their own coat proteins, and satellite RNAs (or "virusoids"), which use the coat protein of the helper virus (Appendix 2). Satellites replicate in the cytoplasm using an RNA-dependent RNA polymerase, an enzymatic activity found in plant but not animal cells. Typical properties of satellites include:

- Their genomes have approximately 500−2,000 nt of single-stranded RNA.
- Unlike defective virus genomes, there is little or no nucleotide sequence similarity between the satellite and the helper virus genome.

Principles of Molecular Virology. DOI: http://dx.doi.org/10.1016/B978-0-12-801946-7.00008-0

- They cause distinct disease symptoms in plants that are not seen with the helper virus alone.
- Replication of satellites usually interferes with the replication of the helper virus (unlike most defective virus genomes).

Satellites of plant viruses do not interfere with the replication of their helper viruses. In contrast, some recently discovered satellite-like viruses of animal viruses (such as the "Sputnik" virophage) do reduce helper virus production, and have become known as virophages—"eaters of viruses." A number of different virophages have been found and all replicate in cells coinfected with large DNA viruses such as Mimivirus. Inhibition of helper virus replication makes virophages distinct from true satellites, even though their biology more closely resembles satellites than anything else.

Viroids are very small (200–400 nt), rod-like RNA molecules with a high degree of secondary structure (Figure 8.1). They have no capsid or envelope and consist only of a single nucleic acid molecule. Viroids are associated with plant diseases and are distinct from satellites in a number of ways (Table 8.1). The first viroid to be discovered and the best studied is potato spindle tuber viroid (PSTVd; viroid names are abbreviated "Vd" to distinguish them from viruses). Viroids do not encode any proteins and are replicated by host-cell RNA polymerase II or possibly by the product of an RNA-dependent RNA polymerase gene in some eukaryotic cells. The details of replication are not understood, but it is likely to occur by a rolling circle mechanism followed by autocatalytic cleavage and self-ligation to produce the mature viroid.

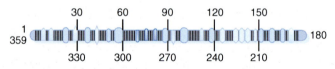

FIGURE 8.1 Structure of a viroid RNA.
Viroids are small (200–400 nt), rod-like RNA molecules with a high degree of secondary structure.

Table 8.1 Satellites and Viroids

Characteristic	Satellites	Viroids
Helper virus required for replication	Yes	No
Protein(s) encoded	Yes	No
Genome replicated by	Helper virus enzymes	Host-cell RNA polymerase II
Site of replication	Same as helper virus (nucleus or cytoplasm)	Nucleus

All viroids share a common structural feature: a conserved central region of the genome believed to be involved in their replication (Figure 8.2). One group of viroids is capable of forming a hammerhead structure, giving them the enzymatic properties of a ribozyme (an autocatalytic, self-cleaving RNA molecule). This activity is used to cleave the multimeric structures produced during the course of replication. Other viroids use unknown host nuclear enzymes to achieve this objective. Some viroids (e.g., cadang-cadang coconut viroid or CCCVd) cause severe and lethal disease in their host plants. Others range from no apparent pathogenic effects (e.g., hop latent viroid or HLVd) to mild disease symptoms (e.g., apple scar skin viroid or ASSVd). It is not clear how viroids cause pathogenic symptoms, but obviously these must result from some perturbation of the normal host-cell metabolism. They show some similarities with certain eukaryotic host-cell sequences, in particular with an intron found between the 5.8S and 25S ribosomal RNAs and with the U3 snRNA which is involved in splicing; therefore, it has been suggested that viroids may interfere with posttranscriptional RNA processing in infected cells. *In vitro* experiments with purified mammalian protein kinase PKR (Chapter 6) have shown that the kinase is strongly activated (phosphorylated) by viroid strains that cause severe symptoms, but far less by mild strains. Activation of a plant homologue of PKR could be the triggering event in viroid pathogenesis (see discussion of the hypersensitive response in Chapter 6).

Most viroids are transmitted by vegetative propagation (i.e., division of infected plants), although a few can be transmitted by insect vectors (nonpropagative) or mechanically. Because viroids do not have the benefit of a protective capsid, viroid RNAs would be expected to be at extreme risk of degradation in the environment; however, their small size and high degree of secondary structure protects them to a large extent, and they are able to persist in the environment for a sufficiently long period to be transferred to another host. The origins of viroids are obscure. One theory is that they may be the most primitive type of RNA genome—possibly leftovers from the "RNA world" believed to have existed during the era of prebiotic evolution.

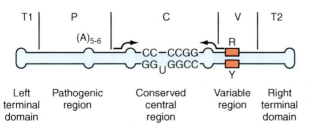

FIGURE 8.2 Functional regions of viroid RNA molecules.
All viroids share common structural features. The central region of the genome is highly conserved and believed to be involved in replication.

Alternatively, they may have evolved at a much more recent time as the most extreme type of parasite. We may never know which of these alternatives is true, but viroids exist and cause disease in plants and in humans.

Hepatitis delta virus (HDV) is a unique chimeric molecule with some of the properties of a **satellite** virus and some of a **viroid** (Table 8.2) which causes disease in humans. HDV requires hepatitis B virus (HBV) as a helper virus for replication and is transmitted by the same means as HBV, benefiting from the presence of a protective coat composed of lipid plus HBV proteins. Virus preparations from HBV/HDV-infected animals contain heterologous particles distinct from those of HBV but with an irregular, ill-defined structure. These particles are composed of HBV antigens and contain the covalently closed circular HDV RNA molecule in a branched or rod-like configuration similar to that of other viroids (Figure 8.3). Unlike all other viroids, HDV encodes a protein, the δ antigen, which is a nuclear phosphoprotein.

Table 8.2 Properties of HDV

Satellite-Like Properties	Viroid-Like Properties
Size and composition of genome—1,640 nt (about four times the size of plant viroids)	Sequence homology to the conserved central region involved in viroid replication
Single-stranded circular RNA molecule	
Dependent on HBV for replication—HDV RNA is packaged into coats consisting of lipids plus HBV encoded proteins	
Encodes a single polypeptide, the δ antigen	

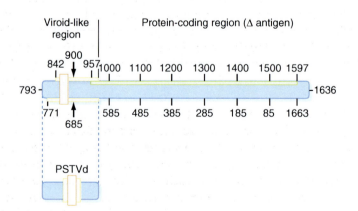

FIGURE 8.3 Structure of hepatitis D virus RNA.
A region at the left end of the genome strongly resembles the RNA of plant viroids such as PSTVd, which is shown for comparison.

Posttranscriptional RNA editing results in the production of two slightly different forms of the protein, δAg-S (195 amino acids), which is necessary for HDV replication, and δAg-L (214 amino acids), which is necessary for the assembly and release of HDV-containing particles. The HDV genome is thought to be replicated by host-cell RNA polymerase II using a "rolling circle" mechanism that produces linear concatemers that must be cleaved for infectivity. The cleavage is carried out by a ribozyme domain present in the HDV RNA, the only known example of a ribozyme in an animal virus genome. HDV is found worldwide wherever HBV infection occurs. The interactions between HBV and HDV are difficult to study, but HDV seems to potentiate the pathogenic effects of HBV infection. Fulminant hepatitis (with a mortality rate of about 80%) is 10 times more common in coinfections than with HBV infection alone. Because HDV requires HBV for replication, it is being controlled by HBV vaccination (Chapter 6).

BOX 8.1 IS IT A BIRD? IS IT A PLANE? NO, IT'S A—WHAT EXACTLY?

Hepatitis delta "virus" is unique—there's nothing else like it. For a start, it's not a virus (capable of independent replication), even though we call it that. It looks very, very like a viroid. Except that it encodes a protein (required for its replication)—and viroids don't do that. So what is it exactly?

HDV is one of the test cases that make virologists glad to be alive. Falling through the cracks between boring bacteria and monotonous mammals, HDV makes us think, and ask big questions about "life."

PRIONS

A particular group of transmissible, chronic, progressive infections of the nervous system show common pathological effects and are invariably fatal. Their pathology is similar to that of amyloid diseases such as Alzheimer's syndrome, and to distinguish them from such noninfectious (endogenous) conditions they are known as transmissible spongiform encephalopathies (TSEs). The earliest record of any TSE dates from several centuries ago, when a disease called scrapie was first observed in sheep (see "TSE in Animals," below). Originally thought to be caused by viruses, the first doubts about the nature of the infectious agent involved in TSEs arose in the 1960s. In 1967, Tikvah Alper was the first to suggest that the agent of scrapie might replicate without nucleic acid, and in 1982 Stanley Prusiner coined the term prion (proteinaceous infectious particle)—which according to Prusiner is pronounced "pree-on." The molecular nature of prions has not been unequivocally proved (see "Molecular Biology of Prions," below), but the evidence that they represent a new phenomenon outside the framework of previous scientific understanding is now generally accepted.

PATHOLOGY OF PRION DISEASES

All prion diseases share a similar underlying pathology, although there are significant differences between various conditions. A number of diseases are characterized by the deposition of abnormal protein deposits in various organs (e.g., kidney, spleen, liver, or brain). These "amyloid" deposits consist of accumulations of various proteins in the form of plaques or fibrils depending on their origin—for example, Alzheimer's disease is characterized by the deposition of plaques and "tangles" composed of β-amyloid protein. None of the "conventional" amyloidoses is an infectious disease, and extensive research has shown that they cannot be transmitted to experimental animals. These diseases result from endogenous errors in metabolism caused by a variety of largely unknown factors. Amyloid deposits appear to be inherently cytotoxic. Although the molecular mechanisms involved in cell death are unclear, it is this effect that gives the "spongiform encephalopathies" their name owing to the characteristic holes in thin sections of affected brain tissue viewed under the microscope; these holes are caused by neuronal loss and gliosis. Deposition of amyloid is the end stage of disease, linking conventional amyloidoses and TSEs and explaining the tissue damage seen in both types of disease, but it does not reveal anything about their underlying causes. Definitive diagnosis of TSE cannot be made on clinical grounds alone and requires demonstration of prion protein (PrP) deposition by immunohistochemical staining of postmortem brain tissue, molecular genetic studies, or experimental transmission to animals, as discussed in the following sections.

TSE IN ANIMALS

A number of TSEs have been observed and intensively investigated in animals. In particular, scrapie is the model for our understanding of human TSEs. Some of these diseases are naturally occurring and have been known about for centuries, whereas others have only been observed more recently and are almost certainly causally related to one another.

Scrapie

First described more than 200 years ago, scrapie is a naturally occurring disease of sheep found in many parts of the world, although it is not universally distributed. Scrapie appears to have originated in Spain and subsequently spread throughout Western Europe. The export of sheep from Britain in the nineteenth century is thought to have helped scrapie spread around the world. Scrapie is primarily a disease of sheep although it can also affect goats. The scrapie agent has been intensively studied and has been experimentally transmitted to laboratory animals many times (see "Molecular

Biology of Prions," below). Infected sheep show severe and progressive neurological symptoms such as abnormal gait; they often repeatedly scrape against fences or posts, a behavior from which the disease takes its name. The incidence of the disease increases with the age of the animals. Some countries, such as Australia and New Zealand, have eliminated scrapie by slaughtering infected sheep and by the imposition of rigorous import controls. Work has shown that the land on which infected sheep graze may retain the condition and infect sheep up to three years later.

The incidence of scrapie in a flock is related to the breed of sheep. Some breeds are relatively resistant to the disease while others are prone to it, indicating genetic control of susceptibility. In recent years the occurrence of TSE in sheep in the United Kingdom closely parallels the incidence of bovine spongiform encephalopathy (BSE) in cattle (Figure 8.4). This is probably due to infection with the BSE agent (to which sheep are known to be sensitive) via infected feed. The natural mode of transmission between sheep is unclear. Lambs of scrapie-infected sheep are more likely to develop the disease, but

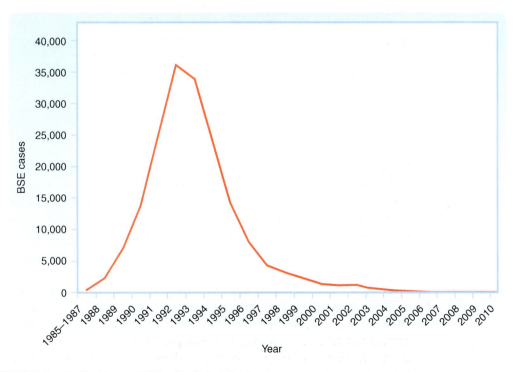

FIGURE 8.4 Reported incidence of BSE in the United Kingdom.
From the earliest recorded incidence in 1985, reported cases of BSE in the United Kingdom have now fallen back to single figures per year. BSE cases continue to be reported from many countries around the world.

the reason for this is unclear. Symptoms of scrapie are not seen in sheep less than one and a half years old which indicates that the incubation period of scrapie is at least this long. The first traces of infectivity can be detected in the tonsils, mesenteric lymph nodes, and intestines of sheep 10−14 months old which suggests an oral route of infection. The infective agent is present in the membranes of the embryo but it has not been demonstrated in colostrum or milk or in tissues of the newborn lambs.

Transmissible Mink Encephalopathy (TME)

TME is a rare disease of farmed mink caused by exposure to a scrapie-like agent in feed. The disease was first identified in Wisconsin in 1947 and has also been recorded in Canada, Finland, Germany, and Russia. Like other TSEs, TME is a slow progressive neurological disease. Early symptoms include changes in habits and cleanliness as well as difficulty in eating or swallowing. TME-infected mink become hyperexcitable and begin arching their tails over their backs, ultimately losing locomotor coordination. Natural TME has a minimum incubation period of 7−12 months, and, although exposure is generally through oral routes, horizontal mink-to-mink transmission cannot be ruled out. The origin of the transmissible agent in TME appears to be contaminated foodstuffs, but this is discussed further below (see "Bovine Spongiform Encephalopathy").

Feline Spongiform Encephalopathy (FSE)

FSE was recognized in the United Kingdom in May 1990 as a scrapie-like syndrome in domestic cats resulting in ataxia (irregular and jerky movements) and other symptoms typical of spongiform encephalopathies. By December 1997, a total of 81 cases had been reported in the United Kingdom. In addition, FSE has been recorded in a domestic cat in Norway and in three species of captive wild cats (cheetah, puma, and ocelot). Inclusion of cattle offal in commercial pet foods was banned in the United Kingdom in 1990, so the incidence of this disease is expected to decline rapidly (see "Bovine Spongiform Encephalopathy").

Chronic Wasting Disease (CWD)

CWD is a disease similar to scrapie which affects deer and captive exotic ungulates (e.g., nyala, oryx, kudu). CWD was first recognized in captive deer and elk in the western United States in 1967 and appears to be endemic in origin. Since its appearance in Colorado, the disease has spread to other states and has also been reported in Canada and South Korea. This disease seems to be more efficiently transmitted from one animal to another than other TSEs, so it seems unlikely that it will ever be eradicated from the regions in which it occurs. CWD prions taken from the brains of infected deer and elk are able to convert normal human prion to a protease-resistant form, a well-studied test for the ability to cause human disease, but the overall risk to human health from this disease remains unclear and there is no evidence that this disease has ever been transmitted to humans.

Bovine Spongiform Encephalopathy

BSE was first recognized in dairy cattle of the United Kingdom in 1986 as a typical spongiform encephalopathy. Affected cattle showed altered behavior and a staggering gait, giving the disease its name in the press of "mad cow disease." On microscopic examination, the brains of affected cattle showed extensive spongiform degeneration. It was concluded that BSE resulted from the use of contaminated foodstuffs. To obtain higher milk yields and growth rates, the nutritional value of feed for farmed animals was routinely boosted by the addition of protein derived from waste meat products and bonemeal (MBM) prepared from animal carcasses, including sheep and cows. This practice was not unique to the United Kingdom but was widely followed in most developed countries. By the end of 2010, a total of 184,607 cases of BSE had been reported in the United Kingdom, and thousands more cases in other countries (Figure 8.4).

The initial explanation for the emergence of BSE in the United Kingdom was as follows. Because scrapie is endemic in Britain, it was assumed that this was the source of the infectious agent in the feed. Traditionally, MBM was prepared by a rendering process involving steam treatment and hydrocarbon extraction, resulting in two products: a protein-rich fraction called "greaves" containing about 1% fat from which MBM was produced and a fat-rich fraction called "tallow" which was put to a variety of industrial uses. In the late 1970s, the price of tallow fell and the use of expensive hydrocarbons in the rendering process was discontinued, producing an MBM product containing about 14% fat in which the infectious material may not have been inactivated. As a result, a ban on the use of ruminant protein in cattle feed was introduced in July 1988 (Figure 8.4). In November 1989, human consumption of specified bovine offals thought most likely to transmit the infection (brain, spleen, thymus, tonsil, and gut) was prohibited. A similar ban on consumption of offals from sheep, goats, and deer was finally announced in July 1996 to counter concerns about transmission of BSE to sheep. The available evidence suggests that milk and dairy products do not contain detectable amounts of the infectious agent. The total number of BSE cases continued to rise, as would be expected from the long incubation period of the disease, and the peak incidence was reached in the last quarter of 1992. Since then the number of new cases has started to fall; however, a number of false assumptions can be identified in the above reasoning.

It is now known that none of the rendering processes used before or after the 1980s completely inactivates the infectivity of prions—therefore cattle would have been exposed to scrapie prions in all countries worldwide where scrapie was present and MBM was used, not just in the United Kingdom in the 1980s. For example, the incidence of scrapie in the United States is difficult to determine, but in the 8 years after the level of compensation for slaughter of infected sheep was raised to $300 in 1977, the reported number of cases went up tenfold to a peak of about 50 affected flocks a year.

BSE is not scrapie. The biological properties of the scrapie and BSE agents are distinct—for example, transmissibility to different animal species and pattern of lesions produced in infected animals (see "Molecular Biology of Prions," below). There is no evidence to support the assumption that BSE is scrapie in cows. The only feasible interpretation based on present knowledge is that BSE originated as an endogenous bovine (cow) prion which was amplified by the feeding of cattle-derived protein in MBM back to cows. Thus, the emergence of BSE in the United Kingdom appears to have been due to a chance event compounded by poor husbandry practices (i.e., use of MBM in ruminant feed).

Important unanswered questions remain concerning BSE. Many of these are raised by the large number of infected cattle born after the 1988 feed ban. It is now generally acknowledged that the feed ban was initially improperly enforced and, moreover, only applied to cattle feed. The same mills that were producing cattle feed were also producing sheep, pig, and poultry food containing MBM, allowing many opportunities for contamination. As a result, in March 1996 the use of all mammalian MBM in animal feed was prohibited in the United Kingdom. It is now known that vertical transmission of BSE in herds can occur at a frequency of 1−10%. Similarly, there is a possibility of environmental transmission similar to that known to occur with scrapie. Apart from the economic damage done by BSE, the main concern remains the possible risk to human health (discussed below).

HUMAN TSEs

There are four recognized human TSEs (summarized in Table 8.3). Our understanding of human TSEs is derived largely from studies of the animal TSEs already described. Human TSEs are believed to originate from three sources:

- **Sporadic:** Creutzfeldt−Jakob disease (CJD) arises spontaneously at a frequency of about one in a million people per year with little variation worldwide. The average age at onset of disease is about 65, and the average duration of illness is about 3 months. This category accounts for 90% of all human TSE, but only about 1% of sporadic CJD cases are transmissible to mice.

- **Iatrogenic (acquired) TSE:** This occurs due to recognized risks (e.g., neurosurgery, transplantation). About 50 cases of TSE were caused in young people who received injections of human growth hormone or gonadotropin derived from pooled cadaver pituitary gland extracts, a practice that has now been discontinued in favor of recombinant DNA-derived hormone.
- **Familial:** Approximately 10% of human TSEs are familial (i.e., inherited). A number of mutations in the human PrP gene are known to give rise to TSE as an autosomal dominant trait acquired by hereditary Mendelian transmission (Figure 8.5).

Table 8.3 TSEs in Human

Disease	Description	Comments
CJD	Spongiform encephalopathy in cerebral and/or cerebellar cortex and/or subcortical gray matter, or encephalopathy with prion protein (PrP) immunoreactivity (plaque and/or diffuse synaptic and/or patchy/perivacuolar types).	Three forms: sporadic, iatrogenic (recognized risk, e.g., neurosurgery), familial (same disease in first degree relative).
Familial fatal insomnia (FFI)	Thalamic degeneration, variable spongiform change in cerebrum.	Occurs in families with PrP_{178} asp-asn mutation.
Gerstmann–Straussler–Scheinker disease (GSS)	Encephalo(myelo)pathy with multicentric PrP plaques.	Occurs in families with dominantly inherited progressive ataxia and/or dementia.
Kuru	Characterized by large amyloid plaques.	Occurs in the Fore population of New Guinea due to ritual cannibalism, now eliminated.

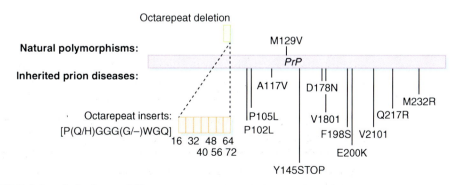

FIGURE 8.5 Mutations in the human PrP gene.
Approximately 10% of human TSEs are inherited. A number of mutations in the human PrP gene are known to give rise to TSEs.

Kuru was the first human spongiform encephalopathy to be investigated in detail and is possibly one of the most fascinating stories to have emerged from any epidemiological investigation. The disease occurred primarily in villages occupied by the Fore tribes in the highlands of New Guinea. The first cases were recorded in the 1950s and involved progressive loss of voluntary neuronal control, followed by death less than 1 year after the onset of symptoms. The key to the origin of the disease was provided by the profile of its victims—it was never seen in young children, rarely in adult men, and was most common in both male and female adolescents and in adult women. The Fore people practiced ritual cannibalism as a rite of mourning for their dead. Women and children participated in these ceremonies but adult men did not take part, explaining the age/sex distribution of the disease. The incubation period for kuru can be in excess of 30 years but in most cases is somewhat shorter. The practice of ritual cannibalism was discouraged in the late 1950s and the incidence of kuru declined dramatically. Kuru has now disappeared.

The above description covers the known picture of human TSEs which has been painstakingly built up over several decades. There is no evidence that any human TSE is traditionally acquired by an oral route (e.g., eating scrapie-infected sheep). There are good reasons why this should be (see "Molecular Biology of Prions," below). However, in April 1996 a paper was published that described a new variant of CJD (vCJD) in the United Kingdom (see Further Reading). Although relatively few in number, these cases shared unusual features that distinguish them from other forms of CJD. The official U.K. Spongiform Encephalopathy Advisory Committee concluded that vCJD is "a previously unrecognized and consistent disease pattern" and that "although there is no direct evidence of a link, on current data and in the absence of any credible alternative the most likely explanation at present is that these cases are linked to exposure to BSE." By 2010, 170 people had died of vCJD in the United Kingdom and several more in other countries. Although three new deaths due to vCJD were recorded in the United Kingdom in 2010, the overall picture is that the vCJD outbreak in the United Kingdom is in decline, albeit with a pronounced tail. One in 2,000 people in the United Kingdom—over 25,000 people—is a carrier of vCJD, for which there is currently still no effective cure. Apart from whether these people will ultimately become ill, another worrying aspect of this statistic is whether these silent carriers can pass the infection on to others through blood transfusions. There is no reliable test to screen for vCJD carriers. Many facts about human prion disease remain unknown, but what is clear is that decades after it started, mad cow disease has not gone away—the effects of the outbreak will rumble on for decades to come.

BOX 8.2 HOW LUCKY DID WE GET?

It seems wrong to say that we got "lucky" with vCJD when 177 people have died in the United Kingdom and nearly 300 have died worldwide. But some of the projections were much worse. At one stage, it was suggested that 14,000 people might die in the United Kingdom alone. As time goes on, the chance of that happening becomes less (but there will still be more deaths for years to come). So is the relatively small number of deaths which actually happened just down to luck? No. It's down to the species barrier which protects each species from the prions of another. Millions of people were exposed to BSE in the United Kingdom alone, but fortunately, humans are quite resistant to cattle prions—more resistant than mice are to hamster prions for example. So it wasn't just luck. But it certainly wasn't down to good planning or judgment.

MOLECULAR BIOLOGY OF PRIONS

The evidence that prions are not conventional viruses is based on the fact that nucleic acid is not necessary for infectivity, as they show:

- Resistance to heat inactivation: Infectivity is reduced but not eliminated by high-temperature autoclaving (135°C for 18 min). Some infectious activity is even retained after treatment at 600°C, suggesting that an inorganic molecular template is capable of nucleating the biological replication of the agent.
- Resistance to radiation damage: Infectivity was found to be resistant to shortwave ultraviolet radiation and to ionizing radiation. These treatments inactivate infectious organisms by causing damage to the genome. There is an inverse relationship between the size of target nucleic acid molecule and the dose of radioactivity or ultraviolet light needed to inactivate them; that is, large molecules are sensitive to much lower doses than are smaller molecules (Figure 8.6). The scrapie agent was found to be highly resistant to both ultraviolet light and ionizing radiation, indicating that any nucleic acid present must be less than 80 nt.
- Resistance to DNAse and RNAse treatment, to psoralens, and to Zn^{2+} catalyzed hydrolysis, all these treatments inactivate nucleic acids.
- Sensitivity to urea, detergents, phenol, and other protein-denaturing chemicals.

All of the above indicate an agent with the properties of a protein rather than a virus. A protein of 254 amino acids (PrP^{Sc}) is associated with scrapie infectivity. Biochemical purification of scrapie infectivity results in preparations highly enriched in PrP^{Sc}, and purification of PrP^{Sc} results in enrichment of scrapie activity. In 1984, Prusiner determined the sequence of 15 amino acids at the end of purified PrP^{Sc}. This led to the discovery that all

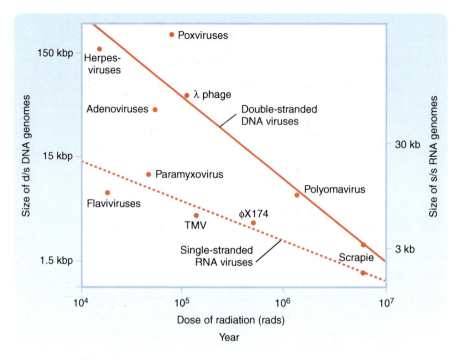

FIGURE 8.6 Radiation sensitivity of infectious agents.
The dose of ionizing radiation required to destroy the infectivity of an infectious agent is dependent on the size of its genome. Larger genomes (e.g., double-stranded DNA viruses, upper line) present a larger "target" and therefore are more sensitive than smaller genomes (e.g., single-stranded RNA viruses, lower line; N.B. φX174 has a single-stranded DNA genome). The scrapie agent is considerably more resistant to radiation than any known virus (note log scale on vertical axis).

mammalian cells contain a gene (*Prnp*) that encodes a protein identical to PrP^{Sc}, termed PrP^C. No biochemical differences between PrP^C and PrP^{Sc} have been determined, although, unlike PrP^C, PrP^{Sc} is partly resistant to protease digestion, resulting in the formation of a 141-amino-acid, protease-resistant fragment that accumulates as fibrils in infected cells (Figure 8.7). Only a proportion of the total PrP in diseased tissue is present as PrP^{Sc}, but this has been shown to be the infectious form of the PrP protein, as highly purified PrP^{Sc} is infectious when used to inoculate experimental animals. Like other infectious agents, there is a dosage effect that gives a correlation between the amount of PrP^{Sc} in an inoculum and the incubation time until the development of disease.

Thus, TSEs, which behave like infectious agents, appear to be caused by an endogenous gene/protein (Figure 8.8). Susceptibility of a host species to prion infection is codetermined by the prion inoculum and the *Prnp* gene.

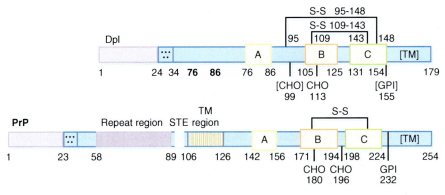

FIGURE 8.7 Structure of the prion (PrP) and Doppel (Dpl) proteins.

The PrP protein and the closely related Doppel (Dpl) protein, overexpression of which also causes neurodegeneration, are part of a family of genes have arisen by gene duplication.

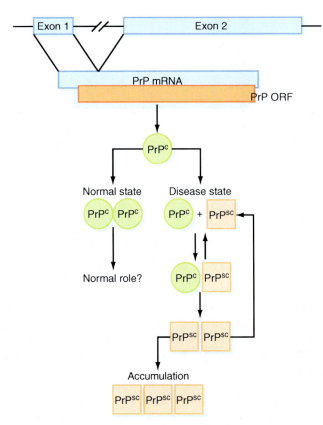

FIGURE 8.8 Schematic diagram of the role of PrP in TSEs.

TSEs behave like infectious agents but appear to be caused by a cellular gene/protein.

Disease incubation times for individual prion isolates vary in different strains of inbred mice, but for a given isolate in a particular strain they are remarkably consistent. These observations have resulted in two important concepts:

1. **Prion strain variation:** At least 15 different strains of PrP^{Sc} have been recognized. These can be determined from each other by the incubation time to the onset of disease and the type and distribution of lesions within the central nervous system (CNS) in inbred strains of mice. Thus, prions can be "fingerprinted," and BSE can be distinguished from scrapie or CJD.

2. **Species barrier:** When prions are initially transmitted from one species to another, disease develops only after a very long incubation period, if at all. On serial passage in the new species, the incubation time often decreases dramatically and then stabilizes. This species barrier can be overcome by introducing a PrP transgene from the prion donor (e.g., hamster) into the recipient mice (Figure 8.9).

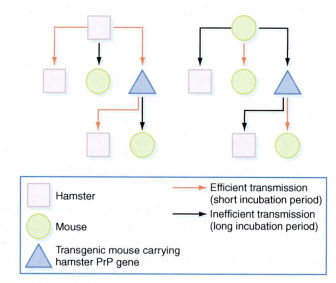

Hamster

Mouse

Transgenic mouse carrying hamster PrP gene

→ Efficient transmission (short incubation period)

→ Inefficient transmission (long incubation period)

FIGURE 8.9 Experimental transmission of scrapie to animals demonstrates a species barrier. Hamster-to-hamster and mouse-to-mouse transmission results in the onset of disease after a relatively short incubation period (75 days and 175 days, respectively). Transmission from one host species to another is much less efficient, and disease occurs only after a much longer incubation period. Transmission of hamster-derived PrP to transgenic mice carrying several copies of the hamster PrP gene (the darker mice in this figure) is much more efficient, whereas transmission of mouse-derived PrP to the transgenic mice is less efficient. Subsequent transmission from the transgenic mice implies that some modification of the properties of the agent seems to have occurred.

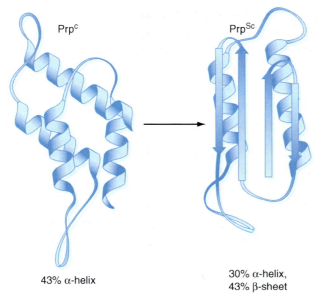

PrP^c

PrP^{Sc}

43% α-helix

30% α-helix,
43% β-sheet

FIGURE 8.10 Conformational changes in PrP.

The fundamental difference between the infectious, pathogenic form (PrP^{Sc}) and the endogenous form (PrP^C) results from a change in the conformation of the folded protein, which adopts a conformation rich in β-sheet.

PrP^C and PrP^{Sc}, however, are not posttranslationally modified, and the genes that encode them are not mutated, which is distinct from Mendelian inheritance of familial forms of CJD. How such apparently complex behavior can be "encoded" by a 254-amino-acid protein has not been firmly established, but there is evidence that the fundamental difference between the infectious, pathogenic form (PrP^{Sc}) and the endogenous form (PrP^C) results from a change in the conformation of the folded protein, which adopts a conformation rich in β-sheet (Figure 8.10).

Transgenic "knockout" Prnp$^{0/0}$ mice that do not possess an endogenous prion gene are completely immune to the effects of PrP^{Sc} and do not propagate infectivity to normal mice, indicating that production of endogenous PrP^C is an essential part of the disease process in TSEs and that the infectious inoculum of PrP^{Sc} does not replicate itself. Unfortunately, these experiments have given few clues to the normal role of PrP^C. Most Prnp$^{0/0}$ mice are developmentally normal and do not have CNS abnormalities, suggesting that loss of normal PrP^C function is not the cause of TSE and that the accumulation of PrP^{Sc} is responsible for disease symptoms. However, one strain of Prnp$^{0/0}$ mouse was found to develop late-onset ataxia and neurological degeneration. This observation led to the discovery of another gene, called Prnd, that is close to the Prnp gene and encodes a 179-amino-acid, PrP-like protein designated Doppel (Dpl), overexpression of which appears to cause neurodegeneration (Figure 8.7). Like *Prnp*, this gene is conserved in vertebrates including humans and may have arisen from *Prnp* by gene duplication. It is suspected that there may be other members of the *Prn* gene family.

The URE3 protein of the yeast *Saccharomyces cerevisiae* has properties very reminiscent of PrP. Other PrP-like proteins are also known (e.g., PSI in yeast and Het-s* in the fungus *Podospora*). URE3 modifies the cellular protein Ure2p, causing altered nitrogen metabolism; similarly, the PSI phenotype involves a self-propagating aggregation of Ure2p and the cellular protein Sup35p. Cells "infected" with URE3 can be "cured" by treatment with protein-denaturing agents such as guanidium, which is believed to cause refolding of URE3 to the Ure2p conformation. The explanation for the inherited familial forms of prion disease is therefore presumably that inherited mutations enhance the rate of spontaneous conversion of PrP^C into PrP^{Sc}, permitting disease manifestation within the lifetime of an affected individual. This concept also suggests that the sporadic incidence of CJD can be accounted for by somatic mutation of the PrP gene and offers a possible explanation for the emergence of BSE—spontaneous mutation of a bovine PrP gene resulted in infectious prions which were then amplified through the food chain. In mammalian cells, the evidence points to prion proteins playing a role in neurotransmission at synaptic junctions. This function of prions as part of the normal epigenetic inheritance of cells indicates their true reason for existence.

> Just as nucleic acids can carry out enzymatic reactions, proteins can be genes.

Reed Wickner

The prion hypothesis is revolutionary and justifiably met with a somewhat skeptical reception. In recent years, the construction of **transgenic** animals has cast further light on the above ideas. PrP is a very difficult protein to work with as it aggregates strongly and is heterogeneous in size, and even the best preparations require 1×10^5 PrP molecules to infect one mouse. This raises the question of whether or not some sort of unidentified infectious agent is "hiding" in the protein aggregates. It is still possible to construct numerous alternative theories of varying degrees of complexity (and plausibility) to fit the experimental data. Science progresses by the construction of experimentally verifiable hypotheses. For many years, research into spongiform encephalopathies has been agonizingly slow because each individual experiment has taken at least one and in some cases many years to complete. With the advent of molecular biology, this has now become a fast-moving and dynamic field. The next few years will undoubtedly continue to reveal more information about the cause of these diseases and will probably

BOX 8.3 ABSENCE OF EVIDENCE OR EVIDENCE OF ABSENCE?

Is there really no nucleic acid associated with prions? And if there isn't, how do you prove that? It took many years for the protein-only prion hypothesis to be generally accepted and for Stanley Prusiner to be awarded the Nobel Prize in 1997. But claims that there may be some sort of nucleic acid or conventional virus associated with prions, even if not a complete genome, just won't go away. How do you prove a negative hypothesis? In the end, scientists have to weigh the balance of evidence and opt for the most likely explanation. And always keep an open mind, just in case.

provide much food for thought about the interaction between infectious agents and the host in the pathogenesis of infectious diseases.

SUMMARY

A variety of novel infectious agents cause disease in plants, in animals, and in humans. Several types of nonviral, subcellular pathogens have disease-causing potential. These include satellites, viroids, and prions. Conventional strategies to combat virus infections, such as drugs and vaccines, have no effect on these unconventional agents. A better understanding of the biology of these novel infectious entities will be necessary before means of treatment for the diseases they cause will become available.

Further Reading

Acevedo-Morantes, C.Y., Wille, H., 2014. The structure of human prions: from biology to structural models—considerations and pitfalls. Viruses 6 (10), 3875–3892.

Aguzzi, A., Zhu, C., 2012. Five questions on prion diseases. PLoS Pathog. 8 (5), e1002651.

Flores, R., Gago-Zachert, S., Serra, P., Sanjuán, R., Elena, S.F., 2014. Viroids: survivors from the RNA world? Annu. Rev. Microbiol. 68 (1), 395–414.

Imran, M., Mahmood, S., 2011. An overview of human prion diseases. Virol. J. 8 (1), 559–567.

Katzourakis, A., Aswad, A., 2014. The origins of giant viruses, virophages and their relatives in host genomes. BMC Biol. 12 (1), 51.

Kovalskaya, N., Hammond, R.W., 2014. Molecular biology of viroid–host interactions and disease control strategies. Plant Sci. 228, 48–60.

Peggion, C., Sorgato, M.C., Bertoli, A., 2014. Prions and prion-like pathogens in neurodegenerative disorders. Pathogens 3 (1), 149–163.

Tseng, C.H., Lai, M.C.C., 2009. Hepatitis delta virus RNA replication. Viruses 1 (3), 818–831.

Appendix 1: Glossary and Abbreviations

Terms shown in the text in bold-colored print are defined in this glossary. A guide to pronunciation is shown in parentheses; this information is intended only as a guide, as there are alternative pronunciations and regional differences in the way many words are pronounced.

abortive infection (a-bore-tiv in-fec-shon) The initiation of infection without completion of the infectious cycle and therefore without the production of infectious particles (cf. productive infection).

adjuvant (aj-oo-vant) A substance included in a medication to improve the action of the other constituents; usually, a component of vaccines that boosts their immunogenicity (e.g., aluminum sulfate).

ambisense (ambi-sense) A single-stranded RNA virus genome that contains genetic information encoded in both the positive (i.e., virus-sense) and negative (i.e., complementary) orientations on the same strand of RNA (e.g., *Bunyaviridae* and *Arenaviridae*) (see Chapter 3).

anergy (an-er-gee) An immunologically unresponsive state in which lymphocytes are present but not functionally active.

apoptosis (ape-oh-toe-sis) The genetically programmed death of certain cells that occurs during various stages in the development of multicellular organisms and may also be involved in control of the immune response.

assembly (ass-embly) A late phase of viral replication during which all the components necessary for the formation of a mature virion collect at a particular site in the cell and the basic structure of the virus particle is formed (see Chapter 4).

attachment (a-tatch-ment) The initial interaction between a virus particle and a cellular receptor molecule; the phase of viral replication during which this occurs (see Chapter 4).

attenuated (at-ten-u-ated) A pathogenic agent that has been genetically altered and displays decreased virulence; attenuated viruses are the basis of live virus vaccines (see Chapter 6).

autocrine (auto-krine) The production by a cell of a growth factor that is required for its own growth; such positive feedback mechanisms may result in cellular transformation (see Chapter 7).

avirulent (a-vir-u-lent) An infectious agent that has *no* disease-causing potential. It is doubtful if such agents really exist—although even the most innocuous organisms may cause disease in certain circumstances (e.g., in immunocompromised hosts).

bacteriophage (back-teer-ee-o-fage) A virus that replicates in a bacterial host cell.

bp (base pair) Base pair—a single pair of nucleotide residue in a double-stranded nucleic acid molecule held together by Watson−Crick hydrogen bonds (see **kbp**).

budding (bud-ding) A mechanism involving release of virus particle from an infected cell by extrusion through a membrane. The site of budding may be at the surface of the cell or may involve the cytoplasmic or nuclear membranes, depending on the site of assembly. Virus **envelopes** are acquired during budding.

capsid (cap-sid) The protective protein coat of a virus particle (see Chapter 2).

chromatin (cro-mat-in) The ordered complex of DNA plus proteins (histones and nonhistone chromosomal proteins) found in the nucleus of **eukaryotic** cells.

cis-acting (sis-acting) A genetic element that affects the activity of contiguous (i.e., on the same nucleic acid molecule) genetic regions; for example, transcriptional promoters and enhancers are *cis*-acting sequences adjacent to the genes whose transcription they control.

complementation (comp-lee-men-tay-shon) The interaction of virus gene products in infected cells that results in the yield of one or both of the parental mutants being enhanced while their genotypes remain unchanged.

conditional lethal mutant (con-dish-on-al lee-thal mu-tant) A conditional mutation whose phenotype is (relatively) unaffected under permissive conditions but which is severely inhibitory under nonpermissive conditions.

conditional mutant (con-dish-on-al mu-tant) A mutant phenotype that is replication competent under "permissive" conditions but not under "restrictive" or "nonpermissive" conditions; for example, a virus with a temperature-sensitive (t.s.) mutation may be able to replicate at the permissive temperature of 33°C but unable to replicate or severely inhibited at the nonpermissive temperature of 38°C.

cytopathic effect (c.p.e.) (sy-toe-path-ik ee-fect) Cellular injury caused by virus infection; the effects of virus infection on cultured cells, visible by microscopic or direct visual examination (see Chapter 7).

defective interfering (D.I.) particles (dee-fect-ive inter-feer-ing part-ik-els) Particles encoded by genetically deleted virus genomes that lack one or more essential functions for replication.

ds (double stranded) Double-stranded (nucleic acid).

eclipse period (ee-clips peer-ee-od) An early phase of infection when virus particles have broken down after penetrating cells, releasing their genomes within the host cell as a prerequisite to replication; often used to refer specifically to bacteriophages (see Chapter 4).

emergent virus (ee-merge-ent vy-rus) A virus identified as the cause of an increasing incidence of disease, possibly as a result of changed environmental or social factors (see Chapter 7).

endemic (en-dem-ik) A pattern of disease that recurs or is commonly found in a particular geographic area (cf. epidemic).

enhancer (en-han-ser) *cis*-Acting genetic elements that potentiate the transcription of genes or translation of mRNAs.

envelope (en-vel-ope) An outer (bounding) lipoprotein bilayer membrane possessed by many viruses. (*Note:* Some viruses contain lipid as part of a complex outer layer, but these are not usually regarded as enveloped unless a bilayer unit membrane structure is clearly demonstrable.)

epidemic (epy-dem-ik) A pattern of disease characterized by a rapid increase in the number of cases occurring and widespread geographical distribution (cf. endemic); an epidemic that encompasses the entire world is known as a pandemic.

eukaryote/eukaryotic (u-kary-ote) An organism whose genetic material is separated from the cytoplasm by a nuclear membrane and divided into discrete chromosomes.

exon (x-on) A region of a gene expressed as protein after the removal of introns by posttranscriptional splicing.

fusion protein (few-shon pro-teen) A virus protein required and responsible for fusion of the virus envelope (or sometimes, the capsid) with a cellular membrane and, consequently, for entry into the cell (see Chapter 4).

genome (gee-nome *or* gen-ome) The nucleic acid comprising the entire genetic information of an organism.

hemagglutination (hay-ma-glut-in-nation) The (specific) agglutination of red blood cells by a virus or other protein.

helix (hee-licks) A cylindrical solid formed by stacking repeated subunits in a constant relationship with respect to their amplitude and pitch (see Chapter 2). (Helical: Shaped like a helix.)

heterozygosis (het-er-o-zy-go-sis) Aberrant packaging of multiple genomes may on occasion result in multiploid particles (i.e., containing more than a single genome) which are therefore heterozygous.

hnRNA (heavy nuclear RNA) "Heterogeneous nuclear RNA" or "heavy nuclear RNA"—the primary, unspliced transcripts found in the nucleus of eukaryotic cells.

hyperplasia (hyper-play-see-a) Excessive cell division or the growth of abnormally large cells; in plants, results in the production of swollen or distorted areas due to the effects of plant viruses.

hypoplasia (high-po-play-see-a) Localized retardation of cell growth. Numerous plant viruses cause this effect, frequently leading to **mosaicism** (the appearance of thinner, yellow areas on the leaves).

icosahedron (eye-cos-a-heed-ron) A solid shape consisting of 20 triangular faces arranged around the surface of a sphere—the basic symmetry of many virus particles (see Chapter 2). (**Icosahedral**: Shaped like an icosahedron.)

immortalized cell (im-mort-al-ized sell) A cell capable of indefinite growth (i.e., number of cell divisions) in culture. On rare occasions, immortalized cells arise spontaneously but are more commonly caused by mutagenesis as a result of virus **transformation** (see Chapter 7).

inclusion bodies (in-klusion bod-ees) Subcellular structures formed as a result of virus infection; often a site of virus assembly (see Chapter 4).

intron (in-tron) A region of a gene removed after transcription by splicing and consequently not expressed as protein (cf. **exon**).

IRES (internal ribosome entry site) (eye-res) An RNA secondary structure found in the 5′ untranslated region (UTR) of (+)sense RNA viruses such as picornaviruses and flaviviruses, which functions as a "ribosome landing pad," allowing internal initiation of translation on the vRNA.

isometric (eye-so-met-rik) A solid displaying **cubic** symmetry, of which the icosahedron is one form.

kb (kilobase) 1,000 nucleotide residues—a unit of measurement of single-stranded nucleic acid molecules; sometimes (wrongly) used to mean **kbp** (below).

kbp (kilobase pair) 1,000 base pairs (see above)—a unit of measurement of double-stranded nucleic acid molecules.

latent period (lay-tent peer-ee-od) The time after infection before the first new extracellular virus particles appear (see Chapter 4).

lysogeny (lie-soj-en-ee) Persistent, latent infection of bacteria by **temperate bacteriophages** such as phage λ.

lytic virus (lit-ik vy-rus) Any virus (or virus infection) that results in the death of infected cells and their physical breakdown.

maturation (mat-yoor-ay-shon) A late phase of virus infection during which newly formed virus particles become infectious; usually involves structural changes in the particle resulting from specific cleavages of capsid proteins to form the mature products, or conformational changes in proteins during assembly (see Chapter 4).

monocistronic (mono-sis-tron-ik) A messenger RNA that consists of the transcript of a single gene and which therefore encodes a single polypeptide; a virus genome that produces such an mRNA (cf. **polycistronic**).

monolayer (mono-layer) A flat, contiguous sheet of adherent cells attached to the solid surface of a culture vessel.

mosaicism (mo-say-iss-cis-em) The appearance of thinner, yellow areas on the leaves of plants caused by the cytopathic effects of plant viruses.

movement protein (move-ment pro-teen) Specialized proteins encoded by plant viruses that modify plasmodesmata (channels that pass through cell walls connecting the cytoplasm of adjacent cells) and cause virus nucleic acids to be transported from one cell to the next, permitting the spread of a virus infection.

mRNA (messenger RNA) Messenger RNA.

multiplicity of infection (m.o.i.) (multi-pliss-itty of in-fect-shon) The (average) number of virus particles that infect each cell in an experiment.

necrosis (neck-ro-sis) Cell death, particularly that caused by an external influence (cf. apoptosis).

negative-sense (neg-at-iv sense) The nucleic acid strand with a base sequence complementary to the strand that contains the protein-coding sequence of nucleotide triplets *or* a virus whose genome consists of a negative-sense strand. (Also "minus-sense" or "(−)sense.")

nonpropagative (non-prop-a-gate-iv) A term describing the transmission via secondary hosts (such as arthropods) of viruses that do not replicate in the vector organism (e.g., geminiviruses). Also known as "noncirculative transmission"; that is, the virus does not circulate in the vector population.

nt (nucleotide) A single nucleotide residue in a nucleic acid molecule.

nucleocapsid (new-clio-cap-sid) An ordered complex of proteins plus the nucleic acid genome of a virus.

oncogene (on-co-gene) A gene that encodes a protein capable of inducing cellular transformation.

ORF (open reading frame) Open reading frame—a region of a gene or mRNA that encodes a polypeptide, bounded by an AUG translation start codon at the 5′ end and a termination codon at the 3′ end. Not to be confused with the poxvirus called orf.

packaging signal (pack-a-jing sig-nal) A region of a virus genome with a particular nucleotide sequence or structure that specifically interacts with a virus protein(s) resulting in the incorporation of the genome into a virus particle.

pandemic (pan-dem-ik) An epidemic that encompasses the entire world.

penetration (pen-ee-tray-shon) The phase of virus replication at which the virus particle or genome enters the host cell (see Chapter 4).

phage (fage) See bacteriophage.

phenotypic mixing (fee-no-tip-ik mix-ing) Individual progeny viruses from a mixed infection that contain structural proteins derived from both parental viruses.

plaque (plak) A localized region in a cell sheet or overlay in which cells have been destroyed or their growth retarded by virus infection.

plaque-forming units (p.f.u.) (plak forming units) A measure of the amount of viable virus present in a virus preparation; includes both free virus particles and infected cells containing infectious particles ("infectious centers").

plasmid (plas-mid) An extrachromosomal genetic element capable of autonomous replication.

polycistronic (poly-sis-tron-ik) A messenger RNA that encodes more than one polypeptide (cf. **monocistronic**).

polyprotein (poly-pro-teen) A large protein that is posttranscriptionally cleaved by proteases to form a series of smaller proteins with differing functions.

positive-sense (pos-it-iv sense) The nucleic acid strand with a base sequence that contains the protein-coding sequence of nucleotide triplets *or* a virus whose genome consists of a positive-sense strand. (Also "plus-sense" or "(+)sense.")

primary cell (pri-mary sell) A cultured cell explanted from an organism that is capable of only a limited number of divisions (cf. **immortalized cell**).

prion (pree-on) A proteinaceous infectious particle, believed to be responsible for transmissible spongiform encephalopathies such as Creutzfeldt–Jakob disease (CJD) or bovine spongiform encephalopathy (BSE) (see Chapter 8).

productive infection (pro-duct-iv in-fect-shon) A "complete" virus infection in which further infectious particles are produced (cf. **abortive infection**).

prokaryote (pro-kary-ote) An organism whose genetic material is not separated from the cytoplasm of the cell by a nuclear membrane.

promoter (pro-mote-er) A *cis-acting* regulatory region upstream of the coding region of a gene that promotes transcription by facilitating the assembly of proteins in transcriptional complexes.

propagative transmission (prop-a-gate-iv trans-mish-on) A term describing the transmission via secondary hosts (such as arthropods) of viruses that are able to replicate in both the primary host and the vector responsible for their transmission (e.g., plant reoviruses). Also known as "circulative transmission" (i.e., the virus circulates in the vector population).

prophage (pro-fage) The lysogenic form of a temperate bacteriophage genome integrated into the genome of the host bacterium.

proteome (pro-tee-ome) The total set of proteins expressed in a cell at a given time.

provirus (pro-vy-rus) The double-stranded DNA form of a retrovirus genome integrated into the **chromatin** of the host cell.

pseudoknot (s'yoo-doh-not) An RNA secondary structure that causes "frame-shifting" during translation, producing a hybrid peptide containing information from an alternative reading frame.

pseudorevertant (s'yoo-doh-re-vert-ant) A virus with an apparently wild-type phenotype but which still contains a mutant genome—may be the result of genetic **suppression**.

pseudotyping (sue-do-type-ing) Where the genome of one virus is completely enclosed within the capsid or, more usually, the envelope of another virus. An extreme form of **phenotypic mixing**.

quasi-equivalence (kwayz-eye-ee-kwiv-al-ense) A principle describing a means of forming a regular solid from irregularly shaped subunits in which subunits in *nearly* the same local environment form *nearly* equivalent bonds with their neighbors (see Chapter 2).

quasispecies (kwayz-eye-spee-sees) A complex mixture of rapidly evolving and competing molecular variants of RNA virus genomes that occur in most populations of RNA viruses.

receptor (ree-sep-tor) A specific molecule on the surface of a cell to which a virus attaches as a preliminary to entering the cell. May consist of proteins or the sugar residues present on glycoproteins or glycolipids in the cell membrane (see Chapter 4).

recombination (ree-com-bin-nation) The physical interaction of virus genomes in a **superinfected** cell resulting in progeny genomes that contain information in nonparental combinations.

release (ree-lease) A late phase of virus infection during which newly formed virus particles leave the cell (see Chapter 4).

replicase (rep-lick-aze) An enzyme responsible for replication of RNA virus genomes (see **transcriptase**).

replicon (rep-lick-on) A nucleic acid molecule containing the information necessary for its own replication; includes both **genomes** and other molecules such as **plasmids** and **satellites**.

retrotransposon (ret-tro-trans-pose-on) A transposable genetic element closely resembling a retrovirus genome, bounded by long terminal repeats (see Chapter 3).

RNAi (RNA interference) (ar-en-ay-eye) A system in cells that helps to control the activity of genes by means of small RNAs which bind to other RNAs and either increase or decrease their activity.

satellites (sat-el-ites) Small RNA molecules (500−2,000 nt) which are dependent on the presence of a helper virus for replication but, unlike defective viruses, show no sequence homology to the helper virus genome. Larger satellite RNAs may encode a protein (cf. **viroids**, **virusoids**).

shutoff (shut-off) A sudden and dramatic cessation of most host-cell macromolecular synthesis which occurs during some virus infections, resulting in cell damage and/or death (see Chapter 7).

splicing (sp-lice-ing) Posttranscriptional modification of primary RNA transcripts that occurs in the nucleus of **eukaryotic** cells during which **introns** are removed and **exons** are joined together to produce cytoplasmic **mRNAs**.

superinfection (super-infect-shon) Infection of a single cell by more than one virus particle, especially two viruses of distinct types, *or* deliberate infection of a cell designed to rescue a mutant virus.

suppression (sup-press-shon) The inhibition of a mutant phenotype by a second suppressor mutation, which may be either in the virus genome or in that of the host cell (see Chapter 3).

syncytium (sin-sit-ee-um) A mass of cytoplasm containing several separate nuclei enclosed in a continuous membrane resulting from the fusion of individual cells. Plural: **syncytia**.

systemic infection (sis-tem-ik infect-shon) An infection involving multiple parts of a multicellular organism.

temperate bacteriophage (temper-ate bac-teer-ee-o-fage) A bacteriophage capable of establishing a **lysogenic** infection (cf. virulent bacteriophage, a bacteriophage that is not capable of establishing a lysogenic infection and always kills the bacteria in which it replicates).

terminal redundancy (ter-minal ree-dun-dance-ee) Repeated sequences present at the ends of a nucleic acid molecule.

titer (titre) (teet-er *or* tight-er) A relative measure of the amount of a substance (e.g., virus or antibody) present in a preparation.

trans-acting (trans-acting) A genetic element encoding a diffusible product which acts on regulatory sites whether or not these are contiguous with the site from which they are produced—for example, proteins that bind to specific sequences present on any stretch of nucleic acid present in a cell, such as transcription factors (cf. *cis*-acting).

transcriptase (trans-crypt-aze) An enzyme, usually packaged into virus particles, responsible for the transcription of RNA virus genomes (see **replicase**).

transfection (trans-fect-shon) Infection of cells mediated by the introduction of nucleic acid rather than by virus particles.

transformation (trans-form-ay-shon) Any change in the morphological, biochemical, or growth parameters of a cell.

transgenic (trans-gene-ik *or* trans-gen-ik) A genetically manipulated eukaryotic organism (animal or plant) that contains additional genetic information from another species. The additional genes may be carried and/or expressed only in the somatic cells of the transgenic organism or in the cells of the germ line, in which case they may be inheritable by any offspring.

transposons (trans-pose-ons) Specific DNA sequences that are able to move from one position in the genome of an organism to another (see Chapter 3).

triangulation number (tri-ang-u-lay-shon num-ber) A numerical factor that defines the symmetry of an icosahedral solid (see Chapter 2).

tropism (trope-ism) The types of tissues or host cells in which a virus is able to replicate.

uncoating (un-coat-ing) A general term for the events that occur after the penetration of a host cell by a virus particle during which the virus capsid is completely or partially removed and the genome is exposed, usually in the form of a nucleoprotein complex (see Chapter 4).

vaccination (vax-sin-ay-shon) The administration of a vaccine.

vaccine (vax-seen) A preparation containing an antigenic molecule or mixture of such molecules designed to elicit an immune response. Virus vaccines can be divided into three basic types: subunit, inactivated, and live vaccines (see Chapter 6).

variolation (var-ee-o-lay-shon) The ancient practice of inoculating immunologically naive individuals with material obtained from smallpox patients—a primitive form of vaccination (see Chapter 1).

virion (vir-ee-on) Morphologically complete (mature) infectious virus particle.

viroid (vy-royd) Autonomously replicating plant pathogens consisting solely of unencapsidated, single-stranded, circular (rod-like) RNAs of 200−400 nt. Viroids do not encode any protein products. Some viroid RNAs have ribozyme activity (self-cleavage) (cf. satellites, virusoids).

virus-attachment protein (vyr-us at-tatch-ment pro-teen) A virus protein responsible for the interaction of a virus particle with a specific cellular receptor molecule.

virusoids (vy-rus-oyds) Small satellite RNAs with a circular, highly base-paired structure similar to viroid; depend on a host virus for replication and encapsidation but do not encode any proteins. All virusoid RNAs studied so far have ribozyme activity (cf. satellites, viroids).

zoonosis (zoo-no-sis) Infection transmitted from an animal to a human. Plural: zoonoses.

Appendix 2: Classification of Subcellular Infectious Agents

Classifying subcellular infectious agents is more complex than it may appear at first sight, and it is appropriate to start with a few definitions:

- *Systematics* is the science of organizing the history of the evolutionary relationships of organisms.
- *Classification* is determining the evolutionary relationships between organisms.
- *Identification* is recognizing the place of an organism in an existing classification scheme, often using dichotomous keys to identify the organism.
- *Taxonomy* (nomenclature) is assigning scientific names according to agreed international scientific rules. The official taxonomic groups (from the largest to the smallest are):
  - *Kingdom* (e.g., animals, plants, bacteria; does not apply to viruses)
  - *Phylum* (e.g., vertebrates; does not apply to viruses)
  - *Class* (group of related orders; does not apply to viruses)
  - *Order* (group of related families)
  - *Family* (group of related genera)
  - *Genus* (group of related species)
  - *Species*, the smallest taxon

The importance of virus identification has been discussed in Chapter 4. Subcellular agents present a particular problem for taxonomists. They are too small to be seen without electron microscopes, but very small changes in molecular structure may give rise to agents with radically different properties. The vast majority of viruses that are known have been studied because they have pathogenic potential for humans, animals, or plants. Therefore, the disease symptoms caused by infection are one criterion used to aid classification. The physical structure of a virus particle can be determined directly (by electron microscopy) or indirectly (by biochemical or serological investigation) and is also used in classification. However, the structure and sequence of the virus genome have increased in importance as molecular biological analysis provides a rapid and sensitive way to detect and differentiate many diverse viruses.

In 1966, the International Committee on Nomenclature of Viruses was established and produced the first unified scheme for virus classification. In 1973, this committee expanded its objectives and renamed itself the International Committee on Taxonomy of Viruses (ICTV). Rules for virus taxonomy have been established, some of which include:

- Latin binomial names (e.g., *Rhabdovirus carpio*) are not used. No person's name should be used in nomenclature. Names should have international meaning.
- A virus name should be meaningful and should consist of as few words as possible. Serial numbers or letters are not acceptable as names.
- A virus species is a polythetic class (i.e., a group whose members always have several properties in common, although no single common attribute is present in all of its members) of viruses that constitute a replicating lineage and occupy a particular ecological niche.
- A genus is a group of virus species sharing common characters. Approval of a new genus is linked to the acceptance of a type species (i.e., a species that displays the typical characteristics on which the genus is based).
- A family is a group of genera with common characters. Approval of a new family is linked to the acceptance of a type genus.

In general terms, groups of related viruses are divided into families whose names end in the suffix "*viridae*" (e.g., *Poxviridae*). In most cases, a higher level of classification than the family has not been established, although six orders (groups of related families) have now been recognized (see Chapter 3). In a few cases, very large families have been subdivided into subfamilies and end in the suffix "*virinae*." Subspecies, strains, isolates, variants, mutants, and artificially created laboratory recombinants are not officially recognized by the ICTV (see Van Regenmortel, M.H.V. (1999). How to write the names of virus species. *Archives of Virology*, 144(5): 1041–1042).

- The names of virus orders, families, subfamilies, genera, and species should be written in italics with the first letter capitalized.
- Other words are not capitalized unless they are proper nouns (e.g., Tobacco mosaic virus, Poliovirus, Murray River encephalitis virus).
- This format should only be used when official taxonomic entities are referred to—it is not possible to centrifuge the species *Poliovirus*, for example, but it is possible to centrifuge poliovirus.
- Italics and capitalization are not used for vernacular usage (e.g., rhinoviruses, cf. the genus *Rhinovirus*), for acronyms (e.g., HIV-1), nor for adjectival forms (e.g., poliovirus replicase).

In 2013 the ICTV formally recognized:

7 Orders
102 Families
and 2,618 Species of viruses

This formal taxonomy is constantly changing, so readers are advised to perform a Google search for "International Committee on Taxonomy of Viruses" where they will be able to find the latest information for themselves.

Appendix 3: The History of Virology

Those who cannot remember the past are condemned to repeat it.

George Santayana

1796: **Edward Jenner** used cowpox to vaccinate against smallpox. Although Jenner is commonly given the credit for vaccination, variolation, the practice of deliberately infecting people with smallpox to protect them from the worst type of the disease, had been practised in China at least 2000 years previously. In 1774, a farmer named Benjamin Jesty had vaccinated his wife and two sons with cowpox taken from the udder of an infected cow and had written about his experience (see 1979). Jenner was the first person to deliberately vaccinate against any infectious disease (i.e., to use a preparation containing an antigenic molecule or mixture of such molecules designed to elicit an immune response).

1885: **Louis Pasteur** experimented with rabies vaccination, using the term *virus* (Latin for 'poison') to describe the agent. Although Pasteur did not discriminate between viruses and other infectious agents, he originated the terms *virus* and *vaccination* (in honour of Jenner) and developed the scientific basis for Jenner's experimental approach to vaccination.

1886: **John Buist** (a Scottish pathologist) stained lymph from skin lesions of a smallpox patient and saw 'elementary bodies' which he thought were the spores of micrococci. These were in fact smallpox virus particles - just large enough to see with the light microscope.

1892: **Dmiti Iwanowski** described the first 'filterable' infectious agent - tobacco mosaic virus (TMV) - smaller than any known bacteria. Iwanowski was the first person to discriminate between viruses and other infectious agents, although he was not fully aware of the significance of this finding.

1898:   **Martinus Beijerinick** extended Iwanowski's work with TMV and formed the first clear concept of the virus *contagium vivum fluidum* - soluble living germ. Beijerinick confirmed and extended Iwanowski's work and was the person who developed the concept of the virus as a distinct entity.

   Freidrich Loeffler and **Paul Frosch** demonstrated that foot-and-mouth disease is caused by such 'filterable' agents. Loeffler and Frosch were the first to prove that viruses could infect animals as well as plants.

1900:   **Walter Reed** demonstrated that yellow fever is spread by mosquitoes. Although Reed did not dwell on the nature of the yellow fever agent, he and his coworkers were the first to show that viruses could be spread by insect vectors such as mosquitoes.

1908:   **Karl Landsteiner** and **Erwin Popper** proved that poliomyelitis is caused by a virus. Landsteiner and Popper were the first to prove that viruses could infect humans as well as animals.

1911:   **Francis Peyton Rous** demonstrated that a virus (Rous sarcoma virus) can cause cancer in chickens (Nobel Prize, 1966; see 1981). Rous was the first person to show that a virus could cause cancer.

1915:   **Frederick Twort** discovered viruses infecting bacteria.

1917:   **Felix d'Herelle** independently discovered viruses of bacteria and coined the term bacteriophage. The discovery of bacteriophages provided an invaluable opportunity to study virus replication at a time prior to the development of tissue culture when the only way to study viruses was by infecting whole organisms.

1935:   **Wendell Stanley** crystallized TMV and showed that it remained infectious (Nobel Prize, 1946). Stanley's work was the first step toward describing the molecular structure of any virus and helped to further illuminate the nature of viruses.

1938:   **Max Theiler** developed a live attenuated vaccine against yellow fever (Nobel Prize, 1951). Theiler's vaccine was so safe and effective that it is still in use today! This work saved millions of lives and set the model for the production of many subsequent vaccines.

1939:   **Emory Ellis** and **Max Delbruck** established the concept of the 'one-step virus growth cycle' essential to the understanding of virus replication (Nobel Prize, 1969). This work laid the basis for the understanding of virus replication - that virus particles do not 'grow' but are instead assembled from preformed components.

1940:   **Helmuth Ruska** used an electron microscope to take the first pictures of virus particles. Along with other physical studies of viruses, direct visualization of virions was an important advance in understanding virus structure.

1941:   **George Hirst** demonstrated that influenza virus agglutinates red blood cells. This was the first rapid, quantitative method of measuring eukaryotic viruses. Now viruses could be counted!

1945:   **Salvador Luria** and **Alfred Hershey** demonstrated that bacteriophages mutate (Nobel Prize, 1969). This work proved that similar genetic mechanisms operate in viruses as in cellular organisms and laid the basis for the understanding of antigenic variation in viruses.

1949:   **John Enders**, **Thomas Weller**, and **Frederick Robbins** were able to grow poliovirus *in vitro* using human tissue culture (Nobel Prize, 1954). This development led to the isolation of many new viruses in tissue culture.

1950:   **André Lwoff**, **Louis Siminovitch**, and **Niels Kjeldgaard** discovered lysogenic bacteriophage in *Bacillus megaterium* irradiated with ultraviolet light and coined the term prophage (Nobel Prize, 1965). Although the concept of lysogeny had been around since the 1920s, this work clarified the existence of temperate and virulent bacteriophages and led to subsequent studies concerning the control of gene expression in prokaryotes, resulting ultimately in the operon hypothesis of Jacob and Monod.

1952:   **Renato Dulbecco** showed that animal viruses can form plaques in a similar way as bacteriophages (Nobel Prize, 1975). Dulbecco's work allowed rapid quantitation of animal viruses using assays that had only previously been possible with bacteriophages.

   Alfred Hershey and **Martha Chase** demonstrated that DNA was the genetic material of a bacteriophage. Although the initial evidence for DNA as the molecular basis of genetic inheritance was discovered using a bacteriophage, this principle of course applies to all cellular organisms (although not all viruses!).

1957:   **Heinz Fraenkel-Conrat** and **R.C. Williams** showed that when mixtures of purified tobacco mosaic virus (TMV) RNA and coat protein were incubated together virus particles formed spontaneously. The discovery that virus particles could form spontaneously from purified subunits without any extraneous information indicated that the particle was in the free energy minimum state and was therefore the favoured structure of the components. This stability is an important feature of virus particles.

   Alick Isaacs and **Jean Lindemann** discovered interferon. Although the initial hopes for interferons as broad-spectrum antiviral agents equivalent to antibiotics have faded, interferons were the first cytokines to be studied in detail.

Carleton Gajdusek proposed that a 'slow virus' is responsible for the **prion** disease kuru (Nobel Prize, 1976; see 1982). Gajdusek showed that the course of the kuru is similar to that of scrapie, that kuru can be transmitted to chimpanzees, and that the agent responsible is an atypical virus.

1961:   **Sydney Brenner**, **Francois Jacob**, and **Matthew Meselson** demonstrated that bacteriophage T4 uses host-cell ribosomes to direct virus protein synthesis. This discovery revealed the fundamental molecular mechanism of protein translation.

1963:   **Baruch Blumberg** discovered hepatitis B virus (HBV) (Nobel Prize, 1976). Blumberg went on to develop the first vaccine against HBV, considered by some to be the first vaccine against cancer because of the strong association of hepatitis B with liver cancer.

1967:   **Mark Ptashne** isolated and studied the λ repressor protein. Repressor proteins as regulatory molecules were first postulated by Jacob and Monod. Together with Walter Gilbert's work on the *Escherichia coli* Lac repressor protein, Ptashne's work illustrated how repressor proteins are a key element of gene regulation and control the reactions of genes to environmental signals.

   Theodor Diener discovered **viroids**, agents of plant disease that have no protein capsid. Viroids are infectious agents consisting of a low-molecular-weight RNA that contains no protein capsid responsible for many plant diseases.

1970:   **Howard Temin** and **David Baltimore** independently discovered reverse transcriptase in retroviruses (Nobel Prize, 1975). The discovery of reverse transcription established a pathway for genetic information flow from RNA to DNA, refuting the so-called 'central dogma' of molecular biology.

1972:   **Paul Berg** created the first recombinant DNA molecules, circular SV40 DNA genomes containing λ phage genes and the galactose operon of *E. coli* (Nobel prize, 1980). This was the beginning of recombinant DNA technology.

1973:   **Peter Doherty** and **Rolf Zinkernagl** demonstrated the basis of antigenic recognition by the cellular immune system (Nobel Prize, 1996). The demonstration that lymphocytes recognize both virus antigens and major histocompatibility antigens in order to kill virus-infected cells established the specificity of the cellular immune system.

1975:   **Bernard Moss**, **Aaron Shatkin**, and colleagues showed that messenger RNA contains a specific nucleotide cap at its 5' end which affects correct processing during translation. These discoveries in reovirus and vaccinia were subsequently found to apply to cellular mRNAs - a fundamental principle.

1976: **J. Michael Bishop** and **Harold Varmus** determined that the oncogene from Rous sarcoma virus can also be found in the cells of normal animals, including humans (Nobel Prize, 1989). Proto-oncogenes are essential for normal development but can become cancer genes when cellular regulators are damaged or modified (e.g. by virus transduction).

1977: **Richard Roberts**, and independently **Phillip Sharp**, showed that adenovirus genes are interspersed with noncoding segments that do not specify protein structure (introns) (Nobel Prize, 1993). The discovery of gene splicing in adenovirus was subsequently found to apply to cellular genes - a fundamental principle.

Frederick Sanger and colleagues determined the complete sequence of all 5375 nucleotides of the bacteriophage φX174 genome (Nobel Prize, 1980). This was the first complete genome sequence of any organism to be determined.

1979: Smallpox was officially declared to be eradicated by the World Health Organization (WHO). The last naturally occurring case of smallpox was seen in Somalia in 1977. This was the first microbial disease ever to be completely eliminated.

1981: **Yorio Hinuma** and colleagues isolated human T-cell leukaemia virus (HTLV) from patients with adult T-cell leukaemia. Although several viruses are associated with human tumours, HTLV was the first unequivocal human cancer virus to be identified.

1982: **Stanley Prusiner** demonstrated that infectious proteins he called prions cause scrapie, a fatal neurodegenerative disease of sheep (Nobel Prize, 1997). This was the most significant advance in developing an understanding of what were previously called 'slow virus' diseases and are now known as transmissible spongiform encepthalopathies (TSEs).

1983: **Luc Montaigner** and **Robert Gallo** announced the discovery of human immunodeficiency virus (HIV), the causative agent of AIDS. Within only 2 to 3 years since the start of the AIDS epidemic the agent responsible was identified.

1985: U.S. Department of Agriculture (USDA) granted the first ever license to market a genetically modified organism (GMO) - a virus to vaccinate against swine herpes. The first commercial GMO.

1986: **Roger Beachy**, **Rob Fraley**, and colleagues demonstrated that tobacco plants transformed with the gene for the coat protein of tobacco mosaic virus (TMV) are resistant to TMV infection. This work resulted in a better understanding of virus resistance in plants, a major goal of plant breeders for many centuries.

1989: Hepatitis C virus (HCV), the source of most cases of nonA, nonB hepatitis, was definitively identified. This was the first infectious agent to be identified by molecular cloning of the genome rather than by more traditional techniques (see 1994).

1990: First (approved) human gene therapy procedure was carried out on a child with severe combined immune deficiency (SCID), using a retrovirus vector. Although not successful, this was the first attempt to correct human genetic disease.

1993: Nucleotide sequence of the smallpox virus genome was completed (185,578 bp). Initially, it was intended that destruction of remaining laboratory stocks of smallpox virus would be carried out when the complete genome sequence had been determined; however, this decision has now been postponed indefinitely.

1994: **Yuan Chang, Patrick Moore**, and their collaborators identified human herpesvirus 8 (HHV-8), the causative agent of Kaposi's sarcoma. Using a polymerase chain reaction (PCR)-based technique, representational difference analysis, this novel pathogen was identified.

2001: The complete nucleotide sequence of the human genome was published. About 11% of the human genome is composed of retrovirus-like retrotransposons, compared with only about 2.5% of the genome that encodes unique (nonrepeated) genes!

2003: Number of confirmed cases of people living with HIV/AIDS worldwide reached 46 million, and still the AIDS pandemic continued to grow.

The newly discovered *Mimivirus* became the largest known virus, with a diameter of 400 nm and a genome of 1.2 Mbp.

Severe acute respiratory syndrome (SARS) broke out in China and subsequently spread around the world.

2010: The United Nations Food and Agriculture Organisation (FAO) declares rinderpest virus to be globally eradicated.

2011: 30th anniversary of the discovery of AIDS.

2013: Largest ever outbreak of Ebola virus begins in West Africa.

Index

Note: Page numbers followed by "*f*" and "*t*" refer to figures and tables, respectively.

CPI Antony Rowe
Eastbourne, UK
March 20, 2015